TROUT FISHING

AGENTS IN AMERICA

THE MACMILLAN COMPANY

66 FIFTH AVENUE, NEW YORK

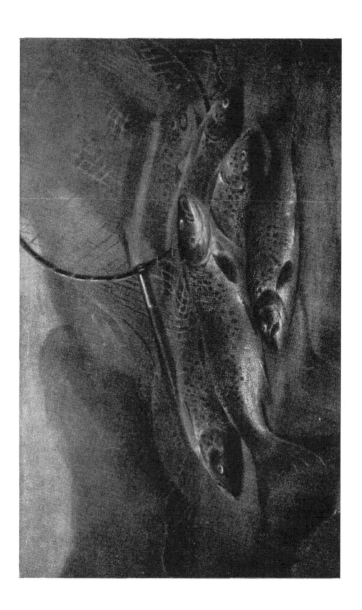

BROWN TROUT.

FROM THE PICTURE BY H. L. ROLFE.

TROUT FISHING

BY

W. EARL HODGSON

WITH A FRONTISPIECE BY H. L. ROLFE AND A FACSIMILE IN
COLOURS OF A MODEL BOOK OF FLIES, FOR STREAM AND
LAKE, ARRANGED ACCORDING TO THE MONTHS
IN WHICH THE LURES ARE APPROPRIATE

LONDON
ADAM AND CHARLES BLACK
1904

PREFACE

THE Book of Flies, inset at the beginning of this volume, is designed for the convenience of those anglers, the great majority, who, amid the pressure of practical affairs, naturally find it difficult to remember the relations of the lures to the months of the season. In arranging the flies for streams I have had the invaluable assistance of Mr. William Senior, who revised, and in some cases added to, the lists which I had drawn up. What are known as "local flies," lures in imitation of insects found only on certain rivers, are not included. Still, it is believed that as regards the flies for running waters the lists are comprehensive. All possible care has been taken to ensure that the images are exactly life-size.

The selection and arrangement of the lake flies has been much more difficult. The few authorities to whom I submitted my own distributions were sceptical as to the possibility of stating exactly what lake flies were appropriate to any particular month. For example, Mr. Robert Anderson, Edinburgh, who has been fishing, and supplying flies to other fishermen, for over forty years, thought that they could be separated only into those which might be called "summer flies" and those which could be used all through the season. This opinion commanded respect; yet there were strong reasons for believing that the very inexact state of the science of lake-fishing was no more than a reflection of the strangely casual manner in which angling is practised on the lakes. These reasons were derived from observation and experience. The insects that flutter about the lakes appear just as regularly, in their seasons, as the insects which haunt the streams;

PREFACE

and they are no less distinct in their
varieties. *It was natural to assume,
therefore, that the flies which would be
fitting lures at one time would not be
fitting at others; and that for the other
times there were appropriate flies, if only
one could find them. The arrangement
set forth in The Book of Flies is the
result of observations and experiments
which have at least been constant and
painstaking.*

*The problem of the lake flies, however,
was not completely solved when the dis-
tribution into months had been settled. In
what sizes were the lures to be presented?
Naturalists admit that the standard sizes
are as a rule larger than the real insects;
yet, in spite of this, practically all anglers
use flies of the standard patterns. This
habit is not in accord with the assumption
set forth in the pages that are to follow,
which is that Nature is the true guide.
Nevertheless, excepting in the cases of the
Green Drake and the Stonefly, which are
life-size, the standards are adopted in The*

b

Book of Flies. After much consideration, there were three reasons for this course. In the first place, however strong might be one's own opinion on the subject of lake flies, which has not until now, I believe, been treated systematically, it seemed right to defer to general usage to the extent, at least, of stating what the usage was. In the second place, experience renders it impossible to deny that sometimes the standard sizes are to be considered right, or, at any rate, not wrong. When the wind is high, all the aspects of a lake even its length and breadth, seem to be on a larger scale, and to grow with the growth of the waves; the very trout increase in voracity and in daring then, and come at the standard flies so well that it is not easy to consider the standards a mistake. In the third place, many of the lakes which contain brown trout contain, at times, sea-trout and salmon also; and in regard to these fish flies larger than the real insects are certainly an advantage. It has been found that salmon now and then, and sea-trout very often, take the lures of which the images are here presented. At the same time, while adopting the standard sizes of the lake flies for these reasons and in deference to usage, I cannot candidly conceal the belief, which is more than theoretical, that even in a high wind lures of smaller size succeed, with the brown trout, just as well; nor ought I to conceal the absolute certainty that in a light wind, or in a calm, lures of the smaller size will be found infinitely better. Indeed, when the wind is light, not only lake flies smaller than the standard, but also some of the stream flies, are often astonishingly successful.

Some may be surprised to see Wasps figuring among the lake flies. Wasps, it may be said, are not water insects. That is true; but neither is the Alder, a favourite on rivers, a water insect in the sense that a Stonefly is. Still, just as the trout in a stream take Alders that are blown on the water by high wind, Wasps sometimes fall upon the lake, and the fish rise at them.

It should, of course, be understood that the lists in The Book of Flies are not to be considered absolutely rigid. As regards weather one month glides into another imperceptibly, and it is not to be supposed that when any month is over all the flies shown under its heading are obsolete for the season. For example, in The Book of Flies the Mayfly appears under the heading "June," because as a rule nature sends it forth in that month, early; but now and then, in the South of England, if the weather is propitious it appears on the streams towards the close of the month after which it is named. Similarly, most of the other insects, like the cereals of the fields, are often a week or two weeks early, or late, according to the weather. The lists in The Book of Flies, then, are to be considered as stating the ascertained averages, not as a code of inflexible time-tables.

Although, if I be not mistaken, The Book of Flies now presented is the first of its kind, pictures of flies, arranged for other purposes, are not uncommon; but much difficulty, I am informed, has been found in the attempts to reproduce the colours exactly. "I warn you," said Mr. Senior, in a letter about my own plan, "that you are likely to have immense trouble over the coloured illustrations; for I have known Halford, Marston, and everybody who has gone through the ordeal, driven frantic in their efforts to get the colours right." Within recent months, happily, there has been much progress in the methods of reproducing coloured pictures; and I am confident that the effort in this volume will be found successful. Through the influence of the publishers, Messrs. A. and C. Black, who have taken a kindly and very gratifying interest in this book, sparing no expense of trouble or of money in its production, I have had high good fortune in the difficulty to which Mr. Senior refers. The artist of The Book of Flies is Mr. Mortimer Menpes. Luck did not end there. On the publishers suggesting that a fron-

tispiece would be acceptable, I remembered
a captivating picture, by Rolfe, hanging
over the fireplace in the hall of a mirthful
shooting-lodge in Kent.   Leave to have
that picture reproduced in " Trout Fish-
ing" was given by the owner, Mr. T. J.
Barratt, willingly.   Indeed, the friendli-
ness of all who have helped me in this
book is so enthusiastic that now I have a
very real apprehension lest the essay itself
should fall short of their expectations.
Among these friends I include Messrs.
Hardy, Alnwick, who made for me the
models of the stream flies, and Mr. Robert
Anderson, who made those of the lake
flies.

It may be that readers of the little book
will now and then seem to catch an echo
of something they have heard or read
before.   If so, that will be because, in the
later days of Mr. Richard Holt Hutton,
I had the honour to write a good many
articles on Angling in " The Spectator,"
and, afterwards, others in " The National
Review," " The Saturday Review," " The

Speaker," " The Academy," " The Daily Chronicle," " The Morning Post," and " The Pall Mall Gazette." It is possible that there may be an echo, or what appears to be one; but that will be merely incidental. This writing as a whole is new. The closing chapter appeared in the " Cornhill Magazine," and part of " The Wind" in " The Daily Mail"; but these were written as integral portions of the book, which, whatever its defects, is the result of an orderly plan.

CONTENTS

xvii

CHAPTER VII

CHAPTER VIII

ILLUSTRATIONS

BROWN TROUT

From the picture by H. L. Rolfe in the possession of
Thomas J. Barratt, Esq., London.

Frontispiece.

A MODEL BOOK OF FLIES FOR STREAM
AND LAKE

Arranged according to the months in which the lures
are appropriate.   Reproduced in facsimile.

Following page xxii.

_{*} *The flies presented in this volume were reproduced
direct at the Menpes Press under the supervision of
Mr. Mortimer Menpes.*

xix

THE BOOK OF FLIES

MARCH FLIES.

STREAM FLIES.

1. GREENWELL'S GLORY.   2. BLUE DUN.   3. OLIVE DUN.

4. FEBRUARY RED.   5. NEEDLE BROWN.   6. BLACK PALMER.

7. RED PALMER.

8. MARCH BROWN (MALE).   9. MARCH BROWN (FEMALE).

10. MARCH BROWN SPIDER.   11. BLAE AND BLACK.   12. MARLOW BUZZ.

13. COW DUNG.   14. WOODCOCK AND HARE'S EAR.

LAKE FLIES.

1. FEBRUARY RED.   2. MARCH BROWN.   3. GROUSE AND CLARET.

4. TEAL AND RED.   5. GREENWELL'S GLORY.   6. HARDY'S FAVOURITE.

⁎ THE FLIES ARE NUMBERED FROM LEFT TO RIGHT.

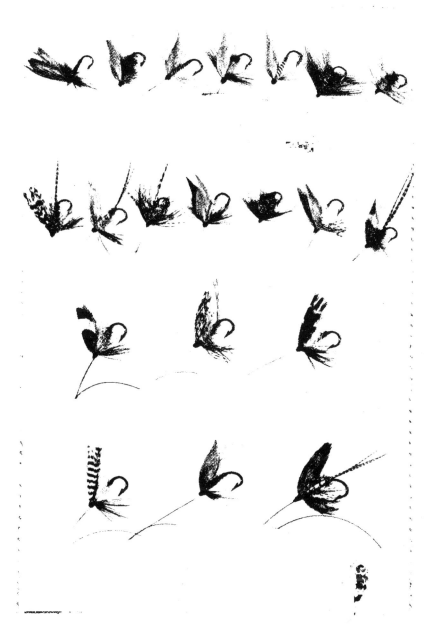

.APRIL FLIES.

STREAM FLIES.

1. RED SPINNER. 2. MARCH BROWN (FEMALE). 3. COW DUNG.

4. LIGHT PARTRIDGE AND YELLOW. 5. WOODCOCK AND ORANGE.

6. BLUE DUN. 7. GOVERNOR. 8. OLIVE DUN.

9. HAWTHORNE FLY. 10. MAY DUN. 11. SAND FLY.

12. WICKHAM'S FANCY. 13. IRON BLUE DUN. 14. RED SPIDER.

15. GRAVEL BED. 16. MARCH BROWN SPIDER. 17. GRANNOM.

LAKE FLIES.

1. LORD SALTOUN. 2. WOODCOCK AND RED. 3. ZULU.

4. MARCH BROWN.

5. BUTCHER. 6. GREENWELL'S GLORY. 7. WOODCOCK AND YELLOW.

MAY FLIES.

STREAM FLIES.

1. WHIRLING DUN.   2. STONE FLY.   3. COACHMAN.
4. LIGHT WOODCOCK AND YELLOW.   5. ALDER.
6. DARK WOODCOCK AND ORANGE.   7. SAND FLY.
8. PALE EVENING DUN.   9. DARK PARTRIDGE.

10. OLIVE DUN.   11. GROUSE AND PEACOCK.   12. WILLOW FLY.
13. YELLOW MAY DUN.   14. TEAL DRAKE.   15. JENNY SPINNER.
16. LIGHT PARTRIDGE.   17. BLACK PALMER.   18. BLACK GNAT.

LAKE FLIES.

1. GOVERNOR.   2. CHALLONER.   3. GROUSE AND GREEN.
4. WOODCOCK AND WILLOW.

5. HECKHAM PECKHAM.   6. TEAL AND BLACK.   7. ALEXANDRA.

JUNE FLIES.

STREAM FLIES.

1. BLACK AND BLAE. 2. HOFLAND'S FANCY. 3. BLACK AND SILVER.

4. RED AND SILVER. 5. BLACK SPINNER. 6. ALDER.

7. GREY QUILL GNAT. 8. BLACK QUILL GNAT. 9. RED QUILL GNAT.

10. OAK FLY.

11. WELSHMAN'S BUTTON. 12. LIGHT BROWN SEDGE. 13. WILLOW FLY.

14. BLACK GNAT. 15. MAY FLY. 16. WATER CRICKET.

17. DARK BROWN SEDGE. 18. RED SPIDER.

LAKE FLIES.

1. GOVERNOR. 2. GROUSE AND OLIVE. 3. TEAL AND GREEN.

4. SLATER. 5. OLIVE QUILL.

6. STONE FLY. 7. GREEN DRAKE. 8. MARLOW BUZZ.

JULY FLIES.

STREAM FLIES.

1. RED PALMER. 2. BLACK PALMER. 3. IRON DUN.
4. WOODCOCK AND HARE'S EAR. 5. WOODCOCK AND RED.
6. WOODCOCK AND BLACK. 7. BLAE AND HARE'S EAR.
8. RED ANT. 9. BLACK ANT. 10. JULY DUN. 11. WILLOW FLY.

12. DOTTEREL AND YELLOW. 13. WREN TAIL. 14. SILVER HORNS.
15. WHITE MOTH. 16. BROWN MOTH. 17. SILVER SEDGE.
18. DARK BROWN SEDGE. 19. ORANGE SEDGE. 20. COACHMAN.

LAKE FLIES.

1. BLUE BOTTLE. 2. ORANGE AND YELLOW WASP. 3. SILVER DOCTOR.
4. BLACK AND ORANGE WASP.

5. BLACK AND YELLOW WASP. 6. SOLDIER PALMER. 7. BROWN PALMER.

AUGUST FLIES.

STREAM FLIES.

1. August Dun. 2. Cinnamon Fly. 3. Dun Midge.

4. Prince Charlie. 5. Jenny Spinner. 6. Willow Fly.

7. Black Spider.

8. Orange Bumble. 9. Honey Dun Bumble. 10. Furnace Palmer.

11. Hardy's Favourite. 12. Dark Brown Sedge.

13. Light Brown Sedge.

LAKE FLIES.

1. Zulu. 2. Alexandra. 3. Butcher.

4. Woodcock and Red Hackle. 5. Blae Wing and Black Hackle.

SEPTEMBER FLIES.

STREAM FLIES.

1. RED SPINNER.   2. WOODCOCK AND HARE'S EAR.   3. BLACK GNAT.

4. RED QUILL.   5. OLIVE QUILL.

6. CINNAMON FLY.   7. BLUE UPRIGHT.   8. CAIRN'S FANCY.

9. GREENWELL'S GLORY.   10. BLUE DUN.

LAKE FLIES.

1. GREENWELL'S GLORY.   2. TEAL AND BLACK HACKLE.

3. WOODCOCK AND HARE'S EAR.

4. GROUSE AND CLARET.   5. BUTCHER.   6. BLAE WING AND RED HACKLE.

7. SAND FLY.

TROUT FISHING

CHAPTER I

KINSHIP WITH THE ARTS

Patience: What Kind?—Fishing and Shooting—
Angling Cannot be Forced—Billiards, Bridge,
and Golf—" Keep your Flies on the Water "—A
Magical Last Resource—Some Idiosyncrasies—
C—— B—— S——, Lochleven Boatman, Mr.
William Senior, Oneself, J—— S——, A——
G——, and Lord A—— —Trout's Sense of
Colour—Sir Herbert Maxwell—*The Spectator*
and Mr. Andrew Lang—Why Fish take
Minnows — Ruddy Mayflies — A Reassuring
Theory—Ptarmigan, Red-deer, and other Wild
Creatures—Colour Must be Right to a Shade—
Floating Flies Sometimes a Mistake—Lord
Granby, Sir Edward Grey, Mr. Sydney Buxton
—The Book of Flies—Its Truth to Nature—
The Quality of Beauty—The Mood of Art.

THERE are many persons who, when they
see a man fishing or hear one speaking
about the sport, smile in an indulgent
indifference. " O no : I could not have

the patience!" they say if asked whether
they go fishing now and then. Although
it has probably been familiar for centuries,
this remark is always a fresh surprise. It
suggests the possibility that the same
worthy persons, if, after seeing a fine
picture, or after hearing a great poem,
they were asked, "Do you paint?" or
"Do you ever write a poem?" might
answer, "O no : I could not have the
patience!" Perhaps, as most of us are
aware, from hearsay, that making pictures
and making poetry are artistic works, and
that all achievements in art are, if only
for our own sakes, to be held in rever-
ence, there are not many inconsiderate
enough to speak about pictures and
poetry in that way. Still, the possibility
of some such startling speech is worth
touching upon. It may coax many good
people into readiness to entertain the
proposition, which otherwise might seem
absurd, that angling is a craft having
subjectively much in common with the
arts of literature and painting.

Patience, which so many persons suppose to be the necessary qualification, is certainly required ; but it is not a thoughtless or inactive patience. It is not merely willingness to wait for an hour, or two hours, or a whole day, watching for an indication that the lure has proved attractive. Patience of that kind has but a small part in the sport. The befitting patience is more than a lazy or stoical endurance. It is continually alert. It embraces much more knowledge and a much greater resourcefulness of thought than are commonly imagined. It is a state of mind more complex than that which is necessary to success in any other pursuit on flood or field.

Contrast it, for example, with that in which one goes out to seek grouse. Instead of having to be lured, the birds are waiting to be shot. Approaching the trout is an action much subtler than walking with a gamekeeper to a place where the grouse are resting. On the grouse-moor a single type of cartridge,

that which is charged with No. 5 shot,
serves all the season round; but the
sportsman on the lake or by the river
has many flies, each fly differing from the
others, and his success depends upon his
knowing the two or three which are
appropriate, in colour, in shape, and in
size, to the time of the year, and even to
the hour of the day. Then, though wilder
at some times than at others, winged
game are not by any weather put wholly
beyond one's reach ; but on a lake, or on a
slowly-running river, a dead calm puts
trout very nearly so, and if the calm is
that of the atmosphere before a thunderstorm it is only by preternatural sagacity
that a fish can be made to rise. In
fine, any man who has a straight eye and
a steady hand can become a good shot;
but the straight eye and the steady hand,
equally needed on the lake or by the
stream, are only, as it were, parts of the
mechanical equipment in the art of
angling. In order that they may be
made effective, eye and hand have to be

informed by a code of knowledge and reflection much wider than that which is needed on the moor. Recently, on a Highland loch, James MacCallum, at the oars, expressed this tersely. "Yes," he said: "ye can force shooting; but ye canna' force fishing."

However intimate any man's acquaintance with the habits of trout may be, there comes not infrequently a day on which it proves astonishingly insufficient. The water is in splendid order, the air is volatile, and the lures seem right; but not a trout will rise. This shows how very elementary the science of angling still is. In the British Islands the sport has been a favourite for centuries. By means of rods and lines, books of flies, and cases of minnow-tackle, as well as by oral tradition and literature, instruction in it has been passed on, constantly revised and expanded, from generation to generation; yet there always have been, and apparently there always will be, days on which, even if his life depended on his doing so,

the most expert angler could not, by fair
means, catch a single trout. Often these
days are to all appearance exactly like
days on which the fish rose at the fly
well and the basket was quickly filled ;
but somehow or other knowledge lingers,
the most experienced skill is baffled. It
is not that all the trout are asleep or fast-
ing. Although they will not look at any
of the lures you offer, here and there you
see one rising or "tailing"; or it may be
that a rapidly-moving upheaval of the
water shows that a large old trout is
rushing at a young one. The fish, or
some of them, are obviously not altogether abstinent from food ; but the task
of catching them passes the wit of man.

This may seem discouraging to any
one who thinks of learning the science
and acquiring the art of angling. Such
an one may say to himself, "What is the
use of trying if it is certain that among
the results will be frequent failure?
Clearly, after all, angling does require a
dull and stupid kind of patience." That

is a superficial view. It is natural to
any one who has either never used rod
and line at all, or has done so, in a
casual manner, only when among a
party of sportsmen at some country-
house ; but to the practised fisherman it
will betray a lack of understanding.
Paradoxical as the notion may seem,
much of the fascination of the pursuit
of trout, which never stales, springs from
the knowledge that the pursuit will often
be unsuccessful. Man, when critically
he examines the habits and the interests
of his leisure times, must realise that he
is a being of strange complexity. He
will cheerfully play billiards for an hour
or so after dinner every night from youth
until in old age the cue trembles in his
hand ; but if one of the incidents of
penal servitude were the daily duty of
playing plain against spot until one or
the other was a thousand up the thought
of gaol would acquire a new and har-
rowing horror. If bridge were not a
voluntary dissipation, attractive because

of the vague sense that there is a slight
wickedness in gambling time and cash
away, the card-rooms at the clubs, which
are crowded every afternoon and evening, would always be as much deserted
as Mayfair is between the Twelfth of
August and the opening of Parliament.
One may question whether even golf
would be played so joyously by so many
thousands if it were part of a compulsory
system of physical training for the nation.
Is not the analogy clear? If one could
always be sure of a heavy basket of trout,
one would go, as a boy goes "unwillingly
to school," unexpectant of any happiness,
facing the hours as a day of tedious duty
to be done. For all the entertainment
to be hoped for, one might as well be
setting out to sea to take in the cod and
haddocks hanging on the lines which had
been set the night before.

Angling cajoles the faculty of observation into a state of pleasurable activity
which can be understood only through
experience. Indolent as he seems as he

drifts on the lake, or saunters up the
stream, casting, casting, casting, the
angler has his mind occupied at every
moment. The trout may be down just
then; but who knows when they may
not be up? Certainly not he unless his
flies are constantly testing the humour of
the fish. An old Lochleven boatman is
wont to say, when some novice in the
sport is showing signs of giving up in
despair, "The first rule here, sir, is—Keep
your flees in the waater. Ye'll never
ha'e a fish unless they're there." This
elementary precept is often neglected.
Many a man gives up for an hour or so
when either he cannot raise a trout or
he sees no rises at natural flies. Often
this results in what should be a good day
turning out a bad one. If none of the
flies which you have been using for half
an hour is successful, another set might
be. Perhaps insects are absent from the
water; but at some hour of the day during the season there certainly should be a
hatch in the course of nature. Untimely

cold may have delayed the rise ; but if an artificial fly chances to be of the proper pattern, the trout will probably take it.

This statement is founded on a memorable incident. A friend in London had been promised three brace of trout before breakfast-time next morning. The lake on which they were to be caught had recently been "fishing so well" that the promise had been made with confidence. It proved to have been rash. Three hours of the afternoon passed without the stirring of a fin. The flies had been changed so often that the resources of the tackle-book seemed exhausted. Indeed, only one fly remained, a thing with a khaki-coloured wing and next to nothing on its body, surely an uninviting lure. Still, it might be tried ; and it was tried ; and within two hours and a half the three brace of trout, packed in heather, were being sped southward by The Flying Scotchman. The despised and nearly rejected fly had raised fish after fish almost as quickly as it could be disengaged and cast once more upon the ripples. It was the Sand Fly; and although, the weather being chill, the insect had not appeared, the time was ripe and the trout had been expecting it.

Coming from a person who essays to discourse on Angling, this will seem a confession of ignorance; and so it is. It will be thought that he should have known when the Sand Fly was due; and so he should. Still, he has something to say for himself. The little incident is four years old. Besides, there never has been, and there is not yet, a man who is all-wise in the craft of angling. The most we can hope to do is to enrich our lore by observation and reflection; and to the accomplishment of this purpose unexpected incidents such as that which has just been narrated contribute greatly. At least, they are capable of doing so. They would do so if one remembered them, thought about them, and interpreted them; but some of us consider them "pure flukes," or freaks on the

part of the fish which will never be repeated, and remember other things which
it were well to forget.

As the knowledge that one must have
unsuspected failings of one's own comes
to the modest mind on observing the unconscious lapses of one's friends, a few
instances of this remembering useless
things may be not out of place.

One morning Mr. C—— B—— S——
and I set forth on Loch Dochart. Charlie
is a barrister-at-law, a man of the world
accomplished in all the knowledge and
the graces of the Town. Though I had
never been out fishing with him before, I
had often heard him talk about the sport ;
and that day I expected to witness a fine
and instructive performance. The morning was all that could be desired. A soft
wind was making a constant movement on
the water ; there were light thin clouds,
now dissolving in rain, anon parting as if
to let the sun glance through ; but the
intervals between my friend's trout were
long. At the other end of the boat the

fish were coming quickly enough : what
could be the matter with Charlie? I
looked round to see; and saw. Charlie
was throwing a very long line, which
went out upon the water so gently that
the fall of the flies was not perceptible;
but the instant after, holding the rod in
his right hand, with the left he pulled in
the line, two arm's lengths, as fast as
his arm could move. Involuntarily, I
expressed astonishment. "Teach your
grandmother," he answered. My learned
friend spoke the words good-humouredly;
but they undoubtedly meant that he
knew what he was doing. I did not dare
to say more about Charlie's error; but I
doubt not that it sprang from his having
once hooked a trout when reeling in his
line, or when the flies were out as a boat
was being rowed ashore or towards some
fresh drift. However this may be, that
day Charlie caught a trout only when
one rose at the moment of his flies alight-
ing : he never had a rise during the jerk-
ing process. Trout do occasionally take

a fly which is being pulled through the
water; but artificial motion causes them as
a rule to remain suspiciously aloof. This
explains why one so often has a rise when
"not looking." Even the most careful
angler, if the trout are rising so badly as
to make him anxious, imparts, in his
eagerness, some little action to the flies;
but when he is "not looking" his arm
and his hand are motionless, the flies seem
natural, and a fish takes the risk. The
same theory is applicable to an experience
which must be common to many an
angler who has visited Lochleven. You
cast for an hour without having a rise,
and, handing your rod to the boatman,
begin to rest. Your pipe is hardly aglow
before the boatman is fast in a lusty
trout! This is simply because he has let
the flies lie a few seconds where they
fell. The boatmen on that interesting
water are as a rule clumsy anglers; but
somehow or other all of them with whom
I am acquainted are free from the error
which, with an exaggeration peculiarly

his own, Charlie illustrated on Loch Dochart.

The incident recalls an exception to the rule that flies should not be dragged. One fine June morning Captain L—— and I were fishing in the Great Stour as it flows round "the garden that I love" so charmingly made famous by Mr. Alfred Austin. When it was time to go in to luncheon, at Swinford Old Manor, I had only one trout. My friend had seven splendid fish, nearly a pound each, to lay out before the Poet Laureate's delighted gaze. As Captain L——, I had noticed, had been casting down-stream and making the fly run up against the current by long pulls, this was a surprise; but the explanation, exceedingly instructive, was at hand. "What fly?" asked our host, enthusiastically. "I do not know its name; but here it is," answered the fisherman, taking his rod from a corner in the hall. "Ah!" said Mr. Austin, whose knowledge of the creatures in the woodlands and the streams is remarkably minute, "the Water

Cricket!" Of all the insects of which imitations are to be found in The Book of Flies, the Water Cricket is, I believe, the only one that runs about on the surface of the stream. All the others, as a rule, move only as the current of the water, or that of the air, ordains.

Every angler, it would seem, has a weakness for some particular fly. Whithersoever he goes, he will give it a chance, and he will continue to believe in it despite any temporary failure. A well-known instance is that of Mr. Senior, the admirable Editor of *The Field*, who trusts so firmly in a certain insect that he has, for the purposes of literature, taken its name as his own, and is familiar to all the world as "Redspinner." He understands that the brilliant creature is at home on every running water at all times of the season, and that it is likely on any day to be attractive to the trout. I myself have similar thoughts about Greenwell's Glory, a fly with a name so aggressive that I make haste

with an explanation. The insect is not
green, and is not arrayed in gauds. His
wings are of a dark dun, and the girdle
of gold encircling his black waistcoat is
like an unobtrusive watch-chain such as a
gentleman of taste might wear. When
first I knew Greenwell, his wings were
cocked upwards over his head in a sprightly
manner, like those of a hawk about to
strike. That was in Scotland. Since
then he has, as it were, changed his tailor,
or rather extended his custom ; and when
he comes forth from London his wings
droop, as if he were a hawk at peace.
Still, Greenwell has lost none of his
attractiveness by having adopted a new
style of dress. His conquests among
the trout I attribute to the probability
that he belongs to a family spread
all over the British Islands. He seems
to have relations wheresoever there
is a lake or a trout-stream, and they
seem to be abroad on the waters, rain
or shine, from March till the end of
September.

Mr. Senior, I doubt not, could give a reasonable explanation for his preference, and I have suggested a justification of my own; but these preferences are not bigoted. Serviceable as the Redspinner and Greenwell's Glory are on many occasions, there are times when other flies are better; but this is a concession which most anglers who have fancies are loath to make. Take, for example, my friend J—— S——. He is remarkably nimble with his little greenheart rod and cast of fine gut. Once in a drift of a mile along the north shore of Loch Doine I saw him catch fifteen big trout; he did not miss a single rise, and did not lose a fish. There could be no more workmanlike sport than that; yet J—— S—— is not free from a superstition which must certainly be at times a handicap. He has an ineradicable belief in the Alder and the Bloody Butcher, one or the other of which, if both of them are not, is always on his cast. Each of these flies once chanced to be the fly of the hour when he used it ; and he thinks, mistakenly, that it is always opportune.

Similarly, having once done well on the Wey with a Mellursh's Fancy, Mr. A—— G——, whithersoever his wandering footsteps stray, is inseparable from that odd lure. It has never occurred to him that the habitation of the insect which it represents is local.

His, however, is an error of omission only. Lord A—— is a sportsman of another kind. He does nothing without reflection. In sport, as in Parliament, he has always a reasoned argument for his conduct. Never when I have been out with him on his fine waters, in North Wales, has he brought home so many trout as were to be expected. Although sometimes one or another of his guests has fared much better, he does not seem surprised. Once, resting by the river at mid-day, I looked at the gear he was using. Although the month was July, the only fly on his cast was a March Brown. Now, like the Redspinner and

Greenwell's Glory, the March Brown is a
lure which it is always well to have handy;
but on that particular day the fly most
noticeably on the water was a blue dun.
I mentioned this to my host, and handed
him my tackle-book. "Take it away," he
said; "take it away! I see you have
them all the colours of the rainbow; but
that's nonsense. I never fish with anything but a March Brown." My ex-
pression of astonishment called forth an
arbitrary doctrine. "Why should I?
Don't you see the earth — 'the brown
old earth'—and the river itself, and
the flies dancing about, and the atmosphere when the sun is clouded? They're
all brown! The very trout are brown—
just like partridges, grouse, pheasants,
hares, and all the other game you can
think of. If you pry into things in a
strong light, you'll detect some different
shades, no doubt; but Nature doesn't pry.
Only the electric light does; and that's
an invention of man, not a thing according
to Nature, —although I will say for it that

it brings out Nature's colour, as when it makes the flame of a candle brown beyond a doubt. Let's to work again. The world is brown, I tell you!"

Although he was in a whimsical mood, there was a real idea amid the banter. Few men have studied trout and their ways so scientifically as Sir Herbert Maxwell has, and the theory which Lord A—— stated half in jest is not more surprising than one which Sir Herbert has advanced in seriousness. It is that, if not absolutely colour - blind, salmon and trout do not pay much attention to the difference between one hue and another. As those who have read his interesting writings will remember, he derived this theory from observations on the Tweed. Never having seen a living insect resembling any of the salmon-flies in use, Sir Herbert Maxwell could not quite believe that it mattered whether it was by a Jock Scott, or a Thunder and Lightning, or a fly of any other pattern, that the salmon were tempted. His scepticism was justified by experiments. He caught salmon with flies which in regard to colour repudiated all local traditions. That, however, does not warrant any definite conclusion. As there is no insect in the least resembling a salmon-fly, it seems absurd to suppose that in taking it the fish is thinking of insects at all. There are at least two possibilities. In the first place, it is conceivable that, without knowing what the lure is like, the fish may snap at it in curiosity or in anger. This conjecture, originally broached by *The Spectator* in a discussion with Mr. Andrew Lang, is not obviously untenable. Many observers, among whom is Sir Herbert Maxwell himself, think that salmon take no food after they quit the sea for the fresh water. If that be so, in snapping at the fly the fish cannot be seeking something to eat, and must be acting upon a purely emotional impulse. In the second place, it is conceivable that, while there is no insect resembling a salmon-fly, the lure

may be not a bad image of some other
living thing. Whatever be the hues of
the feathers of which it is composed, regarded by the human eye while held
against rushing water, or dragged through
calm, it is not at all unlike a minnow or
some other fish of the same size. As
these small fish are various in their hues,
perhaps the explanation lies in this general
similitude. That conjecture is not incompatible with the belief that salmon
feed only when in the sea. There is
reason for suspecting that when a fish of
the salmon kind, or a pike, takes a real
minnow impaled on a flight of hooks, or a
manufactured thing resembling a minnow,
the fish is moved less by a desire to eat
than by a desire to kill. That is only my
own opinion; but it has what seems to be
remarkable evidence in its favour. Many
an angler must have noticed that a salmon
or a trout, like a pike, will leave a whole
shoal of minnows undisturbed and rush at
an impaled minnow or at a phantom.
Why is this? My theory is that the

lure, whether it be an impaled minnow or
an artificial bait, looks like a creature
which is dying or in distress : in the first
case it really is so.   Many wild animals
have an instinct to kill the weaker
brethren.   That is why, for example, the
ailing sheep leaves the flock and hides
itself : it would be killed if it did not go
away.   May not the same instinct govern
the actions of fish ?   My belief that it
does seems borne out by the fact, familiar
to anglers, that a small trout which is
hooked is not unlikely to be seized by a
large one.   The large one passes all the
small fish which are fit and free in order
to kill the one whose unwonted motions
show it to be in distress.

After having upset accepted understandings about the salmon, Sir Herbert
Maxwell made experiments among the
trout, and then published heretical speculations.   He had some artificial Mayflies
dyed red, and some dyed scarlet ; cast
them upon streams, such as the Mimram,
the trout in which are spoken of as having

reached the wariest familiarity with the
angler's wiles; and found just as good
sport as he could have hoped for had the
flies been of the greenish-yellow hue.
This was startling news.   It disturbed
many minds beyond all hope of reassurance.   If trout could not tell red from
yellow, or did not care whether a Mayfly
was one or was the other, clearly all the
thought and pains embodied in the mani-
fold treasures of one's fly-book were
wasted, and pride in one's beautiful
possessions must crumble in chagrin.
Why search the Indies and the Far East
for hackles if feathers which would do as
well were to be found in the nearest
poultry-yard?   Indeed, if trout did not
know one colour from another, or paid no
attention to colour at all, was not the
angler's subtlety a delusion, and the sport
reduced to the level of the laborious
handicrafts?

It has taken one a long time to recover from these misgivings; but hope
revives.   The trout that took the red

and the scarlet Mayflies must have been in a state of panic fearlessness. To venture such a thought may at first seem begging the question ; but that is perhaps because, living in water, where we cannot tarry to observe them, trout in some of their moods are beyond our range of knowledge. To say of a fish whose conduct is irregular that he must be off his head seems even more inconsiderate than saying the same thing of a man whose doings are a perplexity. Of this I am conscious ; and it is not upon an irrational suggestion of mere bewilderment that I rely in hoping to explain away the ruddy Mayflies.

Wild animals whose habits we can observe closely and continuously sometimes behave in a manner which at first sight is quite unaccountable. The ptarmigan are so much in dread of man that they stay habitually on the least easily accessible boulders at the mountain tops ; yet if you come upon a covey of them unawares, they do not take the trouble to fly. In summer and autumn the red-deer, which can scent a man two or three miles off, will, the moment they are conscious of his neighbourhood, trot other miles away from him; yet when the snows of winter cover the heather, they will come down into the glens and beg fodder from the farmers. At all times of the year, sparrows, finches, and other such small birds fight shy of man; yet if in winter, when food is scarce, you throw aniseed to them, suddenly, with a wild whirring of wings and other signs of uncontrollable excitement, they will flutter about you, some of them even resting on your shoulders to ask for more.

Why should it be considered absurd to assume that trout may be occasionally capable of a similar departure from their habitual reserve? If they are not, they differ from most other wild animals with whose instincts we have a fairly complete acquaintance; and to assume this would be more flagrantly unnatural than assuming that, in common with animals of other

species, they do sometimes lose their
judgment and discretion. Besides, the
natural assumption, although not quite
consciously, is already made by anglers
generally, and is even expressed in phrases
which, early in June, inevitably reappear
in all the journals of sport. We hear
of "the Mayfly Carnival": what does
"Carnival" denote if not a hilarious outbreak of reckless indulgence ? We hear
also of "the duffer's fortnight": what can
these words mean save that during the
period of the Mayfly the trout are so
abandoned in voracity that the need for
skill in luring them is for the time gone ?
As food for the fish the Mayflies are
extraordinarily stimulating. When they
are thoroughly "up" and fluttering thickly
about the surface of the stream, all the
trout in the water are near the surface,
gobbling ; even the largest fish, which at
ordinary times lie low unseen, shoulder
the youngsters out of the way and
scour about ravening on the delicacies
of the season. Any one who has

witnessed the wonderful excitement in a river during the Mayfly time will readily realise that then the fish will rush at anything which seems alive.

After all, then, as a test of the trout's sense of colour, Sir Herbert Maxwell's experiments are not by any means conclusive. According to general experience, the sense of colour at ordinary times is marvellously acute. Who cannot recall a day on which the trout showed a preference for some fly so marked as to be practically absolute? The fact which is implicit in that question need not be dwelt upon. It is one of the most familiar phenomena of the sport. If the fish are rising at a dark dun, a pale dun will not do. If you have been catching trout after trout on a woodcock with hare's-ear, you may try a woodcock with red hackle in vain. The presence or the absence of a touch of tinsel on a hook often makes all the difference between success and failure. Some days the tinsel is desired; on others it is forbidding.

The same consideration applies to every
fly in the richest stock. Each has its day
or days, its hour or hours ; and to these
times alone is it opportune. There are
dozens of the flies, a few of them made in
imitation of insects found on certain
waters only, most of them for use anywhere in Great Britain and Ireland.
Think, then, of what a range of knowledge is implied in the fitting choice of
lures to be mounted on the cast. Sometimes, by bringing out the ephemeral
creatures in their due season, Nature
helps : you see on the water, or flying
about just above it, the insects which the
lures on the cast should resemble. Sometimes Nature withholds this help : an
untimely frost, or even a less severe lack
of warmth, delays the hatching.

Often, also, Nature plays a prank
which is injurious to the modern doctrine
that floating flies, to be cast over rising
trout, are the only proper lures. Even
on the warmest day of summer, a chill air
is often not far away. It is wandering

about on the hillsides or on the meadows;
or perchance it lurks in some copse by
the side of the stream. In any case, the
myriad family of insects newly born
among the reeds are liable to be caught
in it ; then they are numbed, fall upon the
water, gradually sink a little below the
surface, and are carried down the stream.
The trout take them without breaking
the water. That explains why the Dry-
fly doctrine is far from being of general
application.   It has been fashionable
within the last ten years. Articles without recorded number, and even a few
books, have been written in its praise. It
has received the unqualified approval of
sportsmen so eminent as Mr. Senior and
Lord Granby, together with the modified
approval of Sir Edward Grey and Mr.
Sydney Buxton among many others ; but
it holds a large element of fallacy. Often
most of the flies provided by Nature are
half-drowned. Half-drowned, then, as a
rule, should be the aspect of the lures
offered to the trout by the angler. Other

considerations leading to that conclusion
will be set forth anon.

Contemplating the great variety of the
flies which any first-class maker of tackle
can provide, one is lost in amazement at
the diligence and the skill which have
gone towards equipment for the sport.
Who discovered all the insects which are
figured in these little structures of feather,
fur, tinsel, silk, and steel ? Some of those
to whom the craft is altogether strange
might question whether in nature there
are so many different insects as a well-
stocked book of flies silently affirms.
Noticing the wealth of colour, the differ-
ences of shape, and the minute individu
alities of texture, they might suppose that,
instead of having been content to copy
nature, the makers of tackle had been
inventing things in the hope that novelties
would captivate the trout.   That would
be misjudging.   Even if it be a wondrous
blend of red, black, yellow, green or blue,
and gold, every one of these things has
its living prototype.   The only difference

is that the creatures of nature are even more beautiful, in some cases more brilliant, in others more delicately neutral, than the creatures of man. Undoubtedly, to those who have eyes to see and diligence to seek, Nature will show the realities. Most of them are born in the neighbourhood of lake or stream; some among the reeds, others in the bushes or the overhanging trees, a few on the bed of the water; all of them, as far as one can perceive, though shrikes and swallows do not disdain them, are designed to be food for trout.

Beautiful we have called them and their images. Why are these things beautiful? That is a question in the philosophy of art; and the answer has truth not for the angler only but also for every man or woman who has a sensibility which gives the word beautiful a meaning. There is not anything which is beautiful in itself. A thing of beauty derives its characteristic quality from its relation to consciousness, a present desire or a memory of pleasure. This is the spirit of all the arts. A beautiful picture, whether its subject be a landscape or a human face, is beautiful because it awakes in us either a sense of our own actual or possible happiness, or the memory of a happiness which is gone. So it is with music, which, although some are strangely insensible to its appeal, as strangely strikes in others chords of association that cannot be traced to any source in this life : music, indeed, is sometimes as a miracle among the arts. So it is with literature : in that domain an achievement having the quality of beauty is a composition in which the artist, while in words which live expressing his own mind on a pleasant theme, makes the mind of another vibrate with a consciousness of pleasure which is, or has been, or yet may be, the other's own. All art lives on association of ideas, and joyous emotions recalled in serenity are an enchantment into the mood of art. That is why the contents of the tackle-book are beautiful. They are associated with past delights, and they suggest delights to come. Looking at them, one can in imagination hear the soft swish of the south-west wind among the sedges and inhale the refreshing perfume of the meadows. Indeed, the memory and the hope of angling bestow upon a rural scene in which there is a lake, or through which there flows a river, a charm that it cannot have for those who have not experienced the sport. To these the lake is a fine sheet of water, out of which a fortune could be made by the owner if there were a large town not far off; and the river has potent falls, which, if they were near enough, could be used to produce a new system of electric lighting for the whole of London. To the fisherman thoughts much more bracing are suggested. He notes the character of the stream : how attractive are its alternations of rapids, deeps, and gravelly pools ! He notes that there is a steady breeze upon the lake : perhaps it is well stocked, and how delightful it would be if one were

afloat on it and a well-thrown fly brought
a game fish dashing with a flash through
the wave ! Even though it is to town
and toil again that one is hastening on
the railroad, the scene sensibly cheers the
journey.

CHAPTER II

THE WIND

A Breeze Desirable—West and South Winds Gener-
ally Favourable—Lochleven and Other Excep-
tions Explained — Trout Not Quite "Cold
blooded" — Direction of Wind does Not
Always Determine Temperature—Anti-cyclonic
Draughts—Why do the Trout Keep Down?—
Want of Wind not In Itself the Cause—A
Theory of Breeze and Ripple—Evidences in its
Favour—Thunderstorms—The Freshened Air
—Lake-fishing Not Coarse Work—Dyed Gut a
Mistake—Wind More Important on Lakes than
on Streams—A Resource in Storm.

WHAT of the wind? Is it high, or low,
or moderate? Is it from the west or
from the south? Is there in it a touch
of east or north?

These are the queries of the angler as
he looks out upon the morning of a day
to be spent in pursuit of trout. Saving

that his hope faints if there seems to be
" thunder in the air," the other conditions
of the weather are comparatively unimportant.   What matters it if there be a
little rain ?   A shower now and then is
refreshing to man and fish ; besides, there
will be fair intervals, in which one's
clothes will dry.   Perhaps the sunshine
is oppressive ; but that need not cause
despair, for clouds are likely to come.

The wind is much more serious.   All
anglers agree on that score.   Especially
if it be in a lake that the trout are to be
sought, a breeze is considered necessary.
If there is no wind, the boat will not
drift, and the trout will not rise at
artificial flies.   If there is too much wind,
the drift will be so quick that many a fish
which would rise had it a chance will
be passed over while another is being
played into the landing net.   To most
anglers this exasperating state of affairs is
very familiar.   At the close of a good
day on a lake during a high wind, who
has not felt that it would have been much

better if only the boat could have been stopped whenever a trout came on? Is it not an article of faith that where one fish rises a good many others are probably feeding?

The direction of the wind is quite as important as its force. If it is from the west or from the south, the trout, it is expected, will rise briskly; if it is from the east or from the north, they will either not move at all or come only in single spies. There are, it is true, exceptions to this rule. Any one, for example, who has fished on Lochleven will remember the gillie's encouraging words if it was against an easterly breeze that the boat cast off from the jetty at Kinross. There are other waters on which winds from the same quarter are not found to tell against the sport. These exceptions are easily explained. Lochleven and the rivers and lakes alluded to are all on the east coast; and an east wind is not so cold, so harsh, directly it leaves the North Sea as it is when it has travelled a good way inland.

Throughout the country at large, however, the rule cannot be denied. It is a
west wind, or a south, that the angler
needs. If the breeze is from either of the
other quarters he has but little hope.
Here and there, as if at some aberrant
bidding, a trout may rise; but he knows
that he will ply his lures diligently and
dexterously for an hour at a time without
success.

Why? Why do trout rise in westerly
or southerly weather and lie low when
the movement of the air is from the north
or from the east?

Many anglers will be disposed to
think that the answer is obvious. Some
will say that trout, not being, as is generally supposed, quite without warmth in
their blood, dislike the cold, and, as
human creatures do, keep out of it when
they can. That theory is not persuasive.
It is true that they err who suppose trout
to be "cold-blooded," many a fish being
distinctly warm as you take it with chill
fingers out of the landing-net; but even

in an east wind the spring or summer
temperature of the atmosphere in daytime is almost always higher than that of
the water. If, therefore, during these
seasons trout wanted to feel the warmth,
they would be constantly rising above
the surface.

Other fishermen will offer a solution
of the problem apparently more scientific.
They will say that the fish as a rule take
artificial flies only when there are on
the water real insects of which the lures
are imitations, and that these insects,
which are aquatic, the eggs lying among
the sedges, or among the weeds and
gravel on the bed of the river or of the
lake, are not brought forth until the cold
winds have passed. This doctrine is more
plausible. It is beyond a doubt that the
fish do not as a rule take artificial flies
freely until the insects which the lures
resemble are fluttering about the water
in abundance. There would, for example,
be no hope in offering a Mayfly before
the natural insect was abroad in its

multitudinous   brilliance.   The   trout
would bolt at sight of it.

Still, the scientific theory about the
fish in relation to the winds is not
sufficient. It assumes that, whilst west
winds and south winds are always warm,
east winds and north winds are always
cold ; and the assumption cannot be
granted. In spring and summer the temperature of the atmosphere is often low
when the wind is from the east, or from
the north, during a cyclone ; but during an
anti-cyclone it is always high. In the
latter case, whencesoever it comes, the air
is at least mild ; often, in July or in August,
it is positively oppressive. That is because
the breezes within the radius of an anti-
cyclone are in a sense not what they
seem. The wind which on the Itchen or
on the Test is from the north-east has
not necessarily come across the seas from
the Polar region. It may not even have
come so far as from London. It is a
north-east wind by courtesy ; but it is not
a true wind at all. It is only a draught.

For all the chill that it contains, it might as well be coming from the Solent. It is not preventing the eggs of the insects from being hatched. If one looks carefully, it will be seen that the flies are certainly on the water.

Why, then, are the trout not rising? The question has never been answered satisfactorily. All we know is that even a draught from the east or from the north puts the fish down, and that they are likely to stay down until the setting of the sun. Even then their mood will not change unless the wind faints away under a clear sky. That, fortunately, often happens in summer; and then, during the cool fresh hour between sundown and the dark, the trout usually rise well.

The brisk sport enlivening that hour is so familiar that most of us have no thought of how astonishing it is. It comes when the atmosphere is still. Should it not, then, cause us to revise the understanding that in daytime we must have a breeze and a ripple if the trout are

to come at the flies ?   The accepted belief
is that in a dead calm, especially if the
sun is unclouded, the trout see the gut to
which the flies are attached, become suspicious, and sink superior to the tempta-
tion ; and that when there is a ripple the
gut is invisible and the flies are of natural
aspect.

This belief is apparently so reasonable
that it has never been openly questioned ;
yet, surely, there are considerations which
shake it.   First, there is the fact, just
noted, that the trout come on with avidity
during the placid evening hour.   The
light is not strong at that time ; but it is
very clear.   To the human eye itself the
gut in the water is visible : presumably, it
cannot escape the notice of the trout,
whose vision is acute.   Besides, there is
not much less light during the hour after
sundown than there is during an hour in
the middle of the day when the sky is
covered by thick clouds.   If the fish
ignore the gut in the twilight, they should
ignore it also during the dusk which some-

times falls while the sun is high; yet,
whilst they rise freely in a calm at the
one time, they do not rise at all in a
calm at the other. It would appear,
then, that the ripple is not in itself the
condition of good sport during the daylight.

May it be that the ripple is only a
symptom of the condition? Can it be
that the wind, which causes the ripple,
causes also a state of the water in which
the fish become lively and disposed to
feed? This suggestion may at first be
flouted; but before discarding it anglers
should take note that their craft, though
of great antiquity, is one which has made
extraordinarily little progress. The main
principles of the sport have for centuries
been accepted by generation after generation in unreflective acquiescence. This
is so markedly the case that, although
accustomed to speaking of trout in certain waters as "wary," or "cunning," or
"sophisticated," we who wield the rod
hardly ever suppose that the fish are

subject to moods which are explicably referable to definite conditions.

If they will not rise on a day which seems in all respects perfect, we suppose they are sulking causelessly, and go home without further thought about the matter. That is treating the trout with scant respect. It is not the way in which we treat cattle, whose moods and attitudes are so definitely determined by atmospherical conditions that the skilled observer in the pastures can actually foretell the weather. We forget that, by the action of steam on the carbonates of lime and magnesia, carbonic-acid gas is constantly being generated under the surface of the earth ; that, although most of it escapes into the outer atmosphere, much of the enervating influence frequently rests in still waters ; and that, therefore, far from being less in need of vitalising oxygen than the animals of the land and the air, the trout in many places are normally more in need. May it be that when

they are not rising the fish are inert because the water is in want of freshening ?

The surmise that this may be so
occurred to me on witnessing a suggestive
incident on a Highland loch. Trout were
needed to replenish a hatchery on the hillside. Each, as it was caught, was put
into a pail of water, in which, ere long,
there were half a dozen. By and by it
was noticed that the fish were languish-
ing. Some of them had turned upon
their backs, and were to all appearance
dying. The gamekeeper took the bailing
pan ; filled it with water from the lake ;
and, holding high his hand, plunged the
water through the air into the pail.
Within two minutes all the fish were as
lively as ever. They had been revived
by a fresh supply of oxygen.

Within the knowledge of most anglers
there are certain undisputed phenomena
supporting the theory to illustrate which
I have described that interesting incident.
Trout do not rise when a thunderstorm is
impending. Why ? It cannot be because

they are afraid of the stillness and the
gloom : often they come on freely in the
middle of a dark and silent night. May
it not be that they remain down because,
like the birds of the air, the beasts of the
field, and mankind, fish are made sluggish
by certain conditions of the atmosphere ?
Soon after the storm has come, often
when it is at the height of its rage and
rattle, the trout rush at the flies as reck-
lessly as any fisherman could desire. Of
this, surely, the natural explanation is
that they have been relieved of depression by the change which has been
wrought in the atmosphere by lightning.

Even when there is no " thunder in the
air," the angler, especially if he be on a
lake, where causes and effects are more
broadly manifest than they are on a river
will sometimes have experiences which
point to the same conclusion. The calm
of many a day is modified by puffs of
wind ; but if the trout do not rise in the
hours of calm they do not rise in the
minutes of ripple. The wind has not

been sufficient to refresh the water and
make their humour light.   Sometimes,
too, in a day of storm there are intervals
of lull ; and if the trout rise in the hours
of ruffling they rise equally well when the
lake is smooth.   The refreshment of the
water and the fish has not passed with
the passing of the wind.

Still, it would be wrong to suppose
that the character of the tackle is unimportant.   It is beyond all doubt that fine
gut is needed on still water.   One cannot
be absolutely certain that this is because
the trout actually see the gut if it be not
fine ; but it should be borne in mind that,
apart from the question whether it is
visible, thick gut has at least two objectionable qualities.   It is less pliable than
fine gut, and deprives the flies of the light
and airy motions which they should have.
It carries with it a shower of spray,
which falls upon the water immediately
after the flies and must tend to alarm the
fish.

These are considerations deserving more

4

heed than they usually receive. Most
anglers take it for granted that fishing on
a lake is coarse work compared with fishing on a stream. There seems to be some
reason for that belief. A stream is narrow
and not very deep, and as a rule any part
of it can be reached by a fly as you walk
along the bank ; a lake is wide and deep,
and even when one has fished a whole day
there are great expanses unexplored. It
is natural to feel that fishing on a lake is
angling on a large scale, calling for less
fragile appliances. In one respect this
view is not altogether wrong. The flies
that come out on lakes are in many cases
larger than those which are common on
rivers, and it is right to assume that the
artificial flies for lakes must as a rule be
larger than those which are proper on
streams. In another respect, however,
the view is wrong. On a stream the flies
do not remain where they fall. They move
down, and in moving beyond the radius
of the shower of spray may float over, or
by the side of, a feeding trout. On a lake,

excepting in so far as they are moved by the angler, who as a rule should not move them at all, they do remain where they alight. On a lake, then, when the water is not ruffled by the wind, it is desirable that the gut should be as fine as is compatible with reasonable strength.

In the hope of making the gut invisible, it is often dyed. Some soak it in a solution of logwood; some in ink; some in tea. All these expedients are rather worse than useless. This will be readily realised if you place a strand of dyed gut and a strand of gut undyed in a crystal bowl of water. The dyed gut will be conspicuous; the undyed, being opaque, will be almost invisible. If the bowl were black, or brown, or blue, or inky, the results would be the reverse; but it should not be forgotten that the colour of water looked through from below, as the trout look, is much more nearly the colour of unstained gut than that of any of the dyes.

For reasons which will be set forth in

another chapter, the wind is less important
on a river than it is on a lake.   Here let
it be mentioned that on the lake there can
hardly, in one sense, be too much of it.
Quite a gentle breeze, if continuous, is
often sufficient to bring the trout up ; but,
if they are feeding in earnest, the wind will
not put them down even though it rises
into a gale.   They will rush at a fly in the
trough of a billow which leaves the bottom
of the lake, at some shallow place, almost
uncovered.   The inspiriting nature of this
astonishing discovery is mitigated only by
the difficulties of fishing on a lake when
the wind is very high.   By way of providing against the emergency, some anglers
take out with them a large stone fastened
to a rope ; the stone is to be dropped
overboard when the boat begins to drift
too quickly.   This plan, which is better
than not going out at all, has the disadvantage that a large fish may entangle the
line round the rope, and break off.   There
is considerable reason for believing that
the trout are often in the best of humours

when the storm is at its highest; but the boat at that time is not easily controlled, and, indeed, is frequently blown ashore. Then the only resource is to go to the quarter of the lake from which the wind is coming, and cast into it from the bank. Sometimes the sport is astonishingly good.

CHAPTER III

THE TEMPERATURE

IN one of the preternatural excursions conducted by M. Jules Verne, there was a pleasing and instructive incident. The explorers came upon a lake in arctic regions. According to all known precedent, the water should have borne a thick sheet of ice; but it was quite open.

54

Although the temperature was below zero, there was not so much as a flake or a ray of ice to be seen. Having allowed his companions to gaze for a few moments in wonderment at this spectacle, the leader of the expedition threw a stone into the lake, and produced a spectacle still stranger. As the ripples spread out in a ring round the splash, arrows of young ice darted after and beyond them; with silent rapidity they darted in all directions; as they flew, the spaces between them were filled up by films; and within ten minutes the lake had such an attractive surface that the intrepid adventurers were skating.

The explanation is simple. The lake was surrounded by hills preventing a breath of air from striking it; motionless water does not freeze; the energy of nature was liberated in the agitation caused by the thrown stone.

We do not need to go to either of the Poles for proof that this fable is not absurd. Satisfactory evidence may be found in territories that are already within the Empire. Men of science tell of an experiment for which all that is necessary are a tub, a glass with water in it, a few pounds of snow, and a handful of salt. You mix the snow and the salt in the tub; and they make a thick, briny slush, the temperature of which is much below the point at which water freezes. Then you place the glass in the tub, leaving the rim above the surface of the brine. If the water becomes motionless, it will remain fluid; but in a quarter of an hour, by which time the cold slush has done its work, shake the glass gently, and the water it contains will freeze.

I myself have not been able to succeed with this experiment; but I have twice been a witness of Nature doing strange things with similar materials.

The first occasion was in Fife on a winter morning. The frost during the night had been intense. On getting up and moving about my room, I heard an unfamiliar sound issuing from the earthen-

ware ewer on the washhand-stand.   It
was not unlike the reedy sigh of a steady
quiet wind on the riverside.   *Shw-sh-sh*
may represent it.   The strange sound was
lost in an explosion.   The jug had burst
at the neck.   On examining, I found that
the water had become a solid block of ice.
Here, as in the incident narrated by M.
Verne, the explanation was not far to
seek.   During the night the water had
been chilled below freezing-point ; but it
had remained fluid because it had remained
still.   Shaken by my movements in the
room, it had quickly congealed ; in doing
so it had expanded, and the vessel gave
way where its narrowing-in caused the
ice to have the greatest pressure.

The other occasion was in the Perth-
shire Highlands late in February.   The
journey from London, overnight, had
been tiresome, and it was very refreshing to be in the sunlit clean air of the
mountains.   What was more natural than
that, seeing a trout-rod hanging ready in
the hall, one should think of a stroll with

it after breakfast ? In England you may
not fish at that time of the year ; but in
Scotland, even under the new Act, you
may, for salmon. The loch just outside
my host's door was ice-bound ; but a mile
to the west there was a considerable
stream which would surely be open.
Many a time I had found excellent sport
among the brown trout in that stream ;
and it was just possible that now there
would be rainbows, three or four thousand
of which had a year before been put into
the loch, which the river feeds. Rainbow
trout, I reminded myself, spawn much
later than the native fish : my merry
friend, Mr. Douglas Hall, who has some
fine ponds, with a considerable stream
through them, at Burton Park in Sussex,
had assured me that there they remained
in good condition until the beginning of
March. It would be interesting to learn
how the strangers were faring in the
Highlands.

The knowledge was not to be easily
gained. That became clear when I

reached the bridge across the stream about a hundred yards from where it joins the loch.   Looking down upon the deep pool under the bridge, I saw scarcely any water at all.   The surface was covered with blocks of white ice, apparently thick ; and from the high banks, down the mossy sides of which water had been trickling before the frost, great clusters of huge icicles hung.   The cascade just above, which is in three stages, each about twelve feet high, was still in play ; but the water was small amid the encrusting ice.   How unlike the appearance of the stream in summer or in autumn !   Then it had been a tawny torrent, often with a flow as good as that of the Test.   Now, meandering through the rough masses of snowy ice, it was a blue trickle not much greater than that of an artificial waterfall in a summer garden.

This may seem an odd similitude ; but it is, I think, true.   Grandeur of a wild kind is one aspect of the Highlands ; but it is not the only aspect.   Even in summer

there is about a well-kept estate in that
region a beauty which, in one of its many
moods, almost dwindles into prettiness.
Everything is so clean, and, in the vast
expanses, so tidy, that, when just arrived
from a town bestrewn with dust or mud,
and littered with the vagrant scraps
of waste cast upon the streets by the
community of millions, one wonders what
to do with the match when a cigarette is
lit. To be negligent with it in this stately
place would be like throwing it upon the
floor of a drawing-room. In winter,
which often is until April is well established, this prettiness of the Highlands is
intensified. At that time, save where the
black crags on the hills are too precipitous
to catch the snows, all the towering land
is white : dazzling white, if the sun shines
unclouded, in the daytime ; softly white,
if the frost is holding, with a faint rose
hue on the irregular peaks, as the shade
of the early twilight creeps slowly upward, gray. Also, the Highlands seem
smaller. Surely, though it tops all the

heights around, that hill cannot be the
one up which you toiled, panting, for three
hours, in search of a royal red-deer, only
six months ago ?   Why, it does not seem
much more than a mile from the valley to
the summit !   Surely, too, that depression
which you can just make out on the side
of the neighbouring hill cannot be the
corry by the sides of which one stood in
the drive of the mountain hares ?   Then it
was two miles up : now it looks almost
within a cleek-shot !   So it is with the loch.
Dried up by the frost are all the innumer-
able rills which in summer made tinkling
music, as if of fairy bells, in the tenuous,
trembling air ; and the loch is low, lower
even than it normally is in July, and
almost perceptibly narrower ; one cannot
speak of its length, for both to the east
and to the west it winds far out of sight.
The few streams which survive the grip
of winter are diminished to an even
greater proportion.

The one in which I had hoped to find
the first trout of the year was invisible

just above the cascade : ice-covered from
bank to bank.   However, it was still
awake underneath ; and I remembered
that a mile farther up there was a long
stretch of it nearly flat ; the sun, at noon-
tide now, would be striking full upon it
there ; perhaps it would be open.

It was partly so.   On all the long
stretch there was no place at which the
stream was free from bank to bank.
Everywhere, from the sides, the shapeless
ice protruded ; the blue water in the
middle was tearing past as if it were a
living thing in fear of enemies on both
flanks ; but here and there the stream
seemed to be holding its own in fighting
the frost, and had actually a few yards in
which to breathe.

I cast the flies into one of those open
spaces ; and cast again, again, and again.
What was the matter ?   Had I forgotten
how to throw a fly ?   The line was falling
heavily, not with a splash exactly, but
with an ungainly mark of its whole
length on the swift water, notably the

gut part of it, which should fall unseen ;
at each successive cast the mark became
larger ; unless I was mistaken, the line
was heavier than it should be. I reeled
up, and looked to see what was wrong.

The cast was like a dainty string of
pearls. Apparently it had in some magical
manner threaded its way through hundreds
of precious stones.   There they were ;
fixed, smooth-crystal, dimly glistening in
the sunbeams ; and set upon the opaque
line, from end to end, with a regularity
which the deftest craft could not excel.

They were frozen drops of water.
How had they been formed ? or, rather,
where ?

Sorrowfully when the lake is unruffled
by a breeze, or the stream is smooth, all
of us know that, as has been mentioned,
a cast of other than thin gut carries forward in its flight a shower ; but had these
solid beads of water been formed when the
line was in the air ?   As they did not melt
when bathed in the sunlight, I realised
that the temperature must be low, and it

was possible to think that the drops had
been frozen in the air ; but a subversive
doubt beset me.   Could it be that the
beads, formed and fixed, had been snatched
bodily from the stream itself ?

This thought was incompatible with
the accepted understanding about water.
Many a midnight, walking homeward
from an hour or two of after-dinner
billiards at the Club, my friend Rudolph
Messel, whose scientific knowledge is
honoured in London and Paris and
Berlin, had entertained me with fascinating discourses on the phenomena of
nature.   One night, when the setting-in of
frost was shown by the transfiguration of
Piccadilly from muddy dinginess into a
steely-gray sparkling under the electric
glow, Dr. Messel had dwelt on the fact
that water is the only fluid which expands
in freezing.   If it contracted, instead of
expanding, all living creatures in the lakes
and streams would, he had said, become
extinct.   The settling of the ice would
begin at the bottom ; and when the

whole body of the water was frozen, as it would quickly be, practically remaining so through the winter and far into the months of spring, the creatures could not survive. All the species of them would disappear. In his undogmatic but suggestive way, Dr. Messel had added that the fact of water being an exception to the rule of fluids in relation to frost was one of the most striking evidences of intelligent design in the universe.

I recalled this discourse in contemplating the string of beads which my fly-cast had become. What had occurred to me was that the stream, instead of being water as commonly understood, H_2O with a temperature not below that of the freezing-point, must be actually ice in molecular motion, ice disguised in the normal motion caused by what is known as the law of gravity. This conjecture would be confirmed if anywhere I could find solid water on the bed of the stream.

I found it.

On looking into an open patch a little

below the bridge from which I had started
on the journey northward up the glen, I
saw, in places, the submerged ice formed,
and in others it seemed actually forming.
Large stones at the bottom, stones from
a foot to two feet in diameter, were, on
the sides of them farthest down the stream,
encrusted in ice, which seemed to be
gradually adding to itself upwards, as if
to envelop the whole; and shapeless
masses of half-solid water, like writhing
white jelly-fish, clung to other stones,
shivering at the impact of the blue gush
as it eddied past.

Here was an exception to the exceptional rule by which water when frozen
floats. What did it mean? Had the
forces of nature got beyond the control
of the creative design? For a moment
one was almost tempted to think that
this really might be so, especially when it
was considered that seeming disorders in
the processes of nature are not uncommon,
as when a late snowstorm kills lambs that
were born in their due time, or when

premature hail suddenly devastates the
orchards, undoing the long work of spring
and summer ; but the thought was passing.   Further examination and reflection
suggested a reassuring theory.

When the temperature falls below
freezing-point the water in a pond does
not begin to solidify immediately.   First
a thin layer on the surface is chilled and
sinks ; then the succeeding layer is chilled
and sinks ; then another ; and so on until,
under the influence of the cold, the whole
body of water, or most of it, has been
transfused within itself, and in the process has reached the freezing-point ; then
the forming of ice begins.   Until it does
begin the mean temperature of the pond
is above freezing-point.   That, however,
could not be said of the Highland stream.
Evidence to the contrary was abundant
at every step.   It might be an exaggeration to say that one could actually see
the flanges of ice that protruded from the
banks extending outwards and gradually
narrowing the open space in the middle

of the watercourse; but, although not literally visible, unquestionably there had been that process, which was probably continuing; for the ice could never have formed had the water not been below the temperature at which it becomes solid when still. What would have happened had this process gone on a few days longer? Soon the whole stream would have been frozen over; but there would not long have been a free channel ˙underneath. Each day the sun at noon would have helped the running water to heave up blocks of the ice; gradually accumulating somewhere, these would have weighed down the lowest layer; the stream would have been dammed; distributed over a broad expanse, it would have settled quickly; and all the way downwards from wherever the stoppage began there would soon have been no flow at all, but only a solid seam of ice.

The formation of ice on the bed of the stream was preventing this catastrophe; and so, while seemingly unable to enforce

her own law, Nature was really fulfilling
her design.   The law that water when
frozen is expanded, and so floats, was in-
applicable to the phenomena under review. All the contents of the watercourse,
those which were fluid and those which
were stable, were chilled below the freezing-point ; and one of the contents might
almost as well as any other have been at
the bottom, or in the middle, or on the
top.

> All nature is but art unknown to thee,
> All chance direction which thou canst not see,

until you look with care.   The apparent
breach of the law was explained by the
consideration ·that the water, all of it
so cold that no part was heavier than
another, remained in motion only because
the necessity of falling a thousand feet in
five miles did not allow it to appear in its
true character, which was that of ice.
The secret was revealed whenever it had
something to rest behind, or to cling to :
as when it became frankly ice in the lee of

the stones on its bed, or clustered in beads
on my cast of flies.

Recollection of the effects of temperatures upon the water will help in a study
of the influence of temperatures upon the
trout.   Whilst approaching this subject
with a sense that it is complex, I am not
without hope of being able to present considerations which will divest it of much
mystery.

Often you hear an angler explaining
away an empty basket by saying that the
weather on the water was too " muggy " or
too " close " ; but you never hear him saying that it was too warm.   In his estima-
tion heat in itself is no hindrance to his
efforts : it is only the conditions which
sometimes accompany heat that are a
trouble.   On the other hand, he will often
tell you, without hesitation, that the
weather has been too cold.   Cold, he will
say, puts down the trout.

The proposition, which is usually absolute, made without reference to times
or seasons, is not in accord with experi-

ence. This must speedily be realised by
all whose wanderings in pursuit of trout
extend from the South of England to
the Highlands. The climates of these
places are not the same. In Cornwall,
or in Devonshire, or in Hampshire, a
shower of snow in March is so unusual
as to be noticeable ; in the Highlands,
until the end of April, it is as common
as a shower of rain, and is not a freak
even so late as Whitsuntide. Besides,
fishing in the North begins much earlier
than in the South. From the Thames
to the Test it is not considered sportsmanlike to seek trout until April ; but in
the North they are fair game a month
before that. It is in the Highlands that
this problem of temperature is to be
looked into most scientifically : it is there
the data are most comprehensive.

What, then, do we find in the North ?
Do one's experiences early in the season
afford sanction for the common belief that
the trout are kept down by cold ? They
do not. " Snow brew," admittedly, is un-

favourable.   Anglers do not expect good
baskets from a flood which is the result of
snows quickly melting in a thaw, and undoubtedly the sport is poor.   The explanation, I think, lies mainly in the action of
cold upon the earthworms.   A warm flood,
a flood which comes with spring rain when
the country is free from snow, entices the
worms to the surface of the soil, and
hurries many of them down the hillsides
to the streams, to feed the trout ; but
melting snow chills the earth more than
the snow itself, and " snow brew " on the
hillsides and on the fields causes the worms
to keep to their winter quarters, which are
farther down than a spade goes at a stroke.
A flood of that kind bringing no food into
the streams, the fish are not on the outlook ; and, unless it happens to run up
against the very mouth of one of them, the
angler's worm is unregarded, as a Mayfly
would be in August.   When the melted
snow has been drained off to the sea things
wear a different aspect for the angler.
The temperature may be even lower than

it was when the "snow-brew" floods were out; but that does not matter. The trout will come at the flies. Even if the temperature is such that your fingers and feet are numb, during the first few weeks of the season, when the weather seems to be free from those thundery and other obscure conditions which are a misfortune to the sportsman later in the year, the fish rise well any day and all day. While the water itself is of normal temperature, the temperature of the air is unimportant. The readiness of the trout to rise is not stopped even by a shower of snow.

Very soon, however, there is cause for astonishment. On a running water the sport of one day is pretty much like that of the day before, with the difference that it is sometimes arrested by conditions which, for our present purpose, we will assume to have little direct relation with the temperature, and that its quality increases as the fish gain in strength and agility; but what has come over the lake? Only last week, let us say, this drift by the north shore yielded many trout; but now a rise is rare.

In order to understand the phenomena of sport in lakes, it is desirable that we should first realise that still water differs from running water in an important respect. A stream is of the same temperature all through. It is just as cold, or as warm, on the surface as at the bottom; just as cold, or as warm, at the sides as in the middle. A lake lacks this equality of temperature. Its waters are much less quickly transfused. It is obvious, for example, that if in April there is a sudden freshet from the high lands where snow still lies in drifts and corries, all round the points at which the hill streams enter there will be places where the lake is colder than it is in the middle. There is a still more powerful though less observable cause for inequality of temperatures throughout a lake. We have seen that before ice begins to form on still water the body of the water has to be reduced to freezing-point. Waters that are shallow, therefore, are covered with ice sooner than those which are deep. That is why there is skating in St. James's Park earlier than on the Serpentine. Also it explains why, whilst the Scottish Championship is run for on Lochleven, Lochlomond is almost constantly free from ice. Lochleven is so shallow that it is covered with ice after frost of a few days' duration ; Lochlomond is so deep that long before the process of transfusion has been sufficient the "cold snap" has given way. In a lake inequalities of temperature are produced also by the direct action of the sun. When the sun beats down upon the water the deeps are less quickly warmed than the shallows ; and the shallows on the south side, having no backing of land to retain the heat, are less quickly warmed than those of the north, where, besides striking aslant upon the water itself, the sunbeams beat directly on the banks, by which part of their warmth is caught and thrown upon the lake.

For this reason, at the opening of the season anglers seek their sport along the northern shores. It is a true instinct that guides them thither. At the opening of the season trout are most frequent where the water is least cold. Why, then, it will be asked, does sport along the shore sometimes fall off when spring has advanced a stage ?

The answer will arise on consideration of a characteristic in which water is in harmony with all substances other than explosives such as gunpowder and nitroglycerine and dynamite. Heat does not act upon it so quickly as it acts upon earth, or so slowly as it acts upon wood ; but water is like all of the substances which we meet in the fields or in the woodlands in that its permeation by heat or by cold is gradual. Small bodies of water, which are more quickly heated than large bodies, are also more quickly chilled. By the middle of April the general body of water in a lake is of a sensibly higher temperature than it was in January;

but by that time another change, a
change affecting the habits of the trout,
has come over the conditions of the lake.
Even as the shallows were the first places
to be warmed when the sun waxed early in
the year, they are the first to become cold
when the frosts, as they are wont to do,
return. It is always on the shallows that
the first ice appears. Thus, a "cold snap"
in early spring will cause the temperature
of the shallows to fall below the mean
temperature of the lake. A similar result
comes of another cause, which seems to
have escaped general notice. During a
succession of sunshines early in the year,
the shallows near the shore have day by
day been made warmer than the deeps ;
but all the time that this has been going
on it is the deeps that have been most
surely gaining. Nightly the shallows
have lost most, if not all, the warmth of
the day ; but by night as well as by day
the deeps have been storing nearly all that
they received. They have been retaining
all except the comparatively small portion

which the laws of nature called upon
them to give up during the nights ; while
night by night, at the same instance, the
shallows have had to part with nearly all
the warmth which sank into them during
the day.   Consequently, although the
shallows continue to be warmed day by
day, there comes a time when even in the
full shine of the sun at noon they are
chiller than the deeps.

That time begins about the middle of
April.   It is then that sport along the
shore falls off.   The trout have neither
ceased to feed nor become more wary.
They have simply sought more comfortable quarters in the deeps.

It goes against the grain to be frequently referring to one's own experience,
and in this book I strive to keep such
references as few as may be, making the
narration, as a rule, oblique ; but sometimes the bearing of personal witness is
inevitable.   It seems to be so now.   The
theory about the change in the haunts of
trout which has just been set forth is

derived from daily observation from the opening of a season. Well-filled baskets were the rule all through March and the first half of April, and these were the fruits of fishing either from the banks or from a boat drifting along the banks ; but suddenly this good fortune was at an end. It became as difficult to catch a brace of trout as it had been to catch a score. One morning when, there being no wind, the lake was placid, I noticed that, while a strip of water extending outwards at least thirty yards from the shore was undisturbed by rises, beyond that trout were moving everywhere. Seeing that the fly on the water was a small insect with grayish-white wings and a black body, I put on a cast of midgets, rowed out into the middle of the loch, and had very good sport indeed. The spell was broken. The manner of the breach was rather surprising to myself as well as to the hospitable household with whom I was staying on a holiday. There, as throughout Scotland and England generally, it is a traditional belief that, excepting on lakes which are shallow all over, trout away from the shores cannot be induced to rise at a fly. I had accepted the tradition; but the results of the experiment that calm morning were, in a pleasant sense, disturbing. It had struck me that, as trout in the middle of the loch were rising at real flies, there was no reason for thinking that they would fight shy of artificial ones; and the expectation had been justified.

Still, the experiment was not yet complete. Thinking that my host and hostess, when I spread out before them the produce of the deeps, which were believed to have no produce at all, would say, "O! but you would have caught them along the shore too, if you had fished there : all that has happened is that the trout have come on the rise again," I tried the shore, tried it quite fairly for half an hour, and did not get a single rise.

Out upon the deeps I pulled the boat
once more; and the basket was two or
three pounds heavier when it was time to
return.

My theory about the influence of the
temperatures seemed demonstrated to the
full; but would it always hold? Day
after day for many days I put it to the
test, and the results were not easily interpreted. On the deeps when the weather
was calm a few trout were usually caught,
and often where there was a wind light
enough to cause only a gentle ripple I
had a good many more than a few; but
when the wind was high enough to make
the wavelets break into spray there was
practically no sport at all. On these
days, too, the trout began to come whenever, in its drift, the boat neared the
shore. For a time this was disconcert-
ing; but it was not, I think, inexplicable.
Spring had been moving on; even at
night, the weather had been mild, and
usually in daytime warm; in the course
of their changes, the temperatures of the

6

water in its various parts had either been
approximately equalised, or had been
raised so much that none of them was
too rigorous for the comfort of the trout.
Other things being equal, it is the
shallows that the fish prefer.   We see
this on rivers.   Trout are to be found in
canals deep enough for large ships; but
that is only because they cannot help
themselves.   In a river their preferences
are unmistakable.   It is not in the very
deep channels that the wise man seeks
them.   A considerable pool in the middle
of a river they will not shun, being able,
from that ambush, to see all the living
dainties that come towards them over the
rim of gravel; but in the very deep
channels they are absentees or merely
"passing through."   It is undoubtedly
the shallows that they prefer: the tail of
the rapids, the rapids themselves, and,
in the slowly-moving stretch which is
usually bounded by the dykes of a miller's
dam, the sides, where the mud-banks
shelve upwards among the sedges.   Their

habits in a lake are similar. They tend
towards the deeps when these are the
least uncomfortable parts of the water;
but they prefer the shallows at other
times.

Sometimes, on a midsummer's eve, one
goes out to fish all night; and then,
whether the water be a lake or it be a
stream, an interesting movement by the
trout is invariably noticeable. They may
have been scared from their places in
moderate shallows during the day; but
when night has fallen, and they cannot
see far into the dusk, they congregate in
waters which, in some cases, are hardly
more than enough to cover them.   Often
at that time they come freely at large flies,
and at a black moth as readily as at a
white one.   That is not because they are
then indifferent as to their food.   It is
because colour gradually lapses as the
light wanes.   If you sit in a garden after
sundown, all the hues in it will slowly,
slowly, fade, until the laurels, which were
green in the light, are dark; until a rose

cannot be distinguished from a lily ; until, indeed, there is left only a general blackness. That is not because you cannot see the colours. It is because the colours are not there to see. Colours are light, light in subtle distributions among matter ; and when the light goes, colours also gradually cease to be. That is why in the darkening a black fly is as good as a white one. In the eyes of the trout there is no difference. Each is only a thing which moves, and therefore seems to live, dimly seen. There is a greater wonder to be pondered by the water-side at night. Why are the fish, among which there may be salmon and sea-trout, gathered so closely in the shallow bays ? Is it for warmth ? I do not think so : the deeps, even at midsummer, would be warmer still. I hesitate over my own conjecture ; but it may be given. I think that the fish have come in, out of the current if the water is a stream, to be free from pressure if it is a lake, to bed. There is always a time in any night when

the fish ignore the flies.   They will take
a gentle, or a worm, it is true ; but why ?
A fish snaps at the bait, I think, only
when, chancing to run against him, the
sunken tackle rouses him from sleep.

CHAPTER IV

THE LIGHT

Trout's Eyesight Very Keen—Deer, Grouse, and
Wild Duck Rely on Other Senses—Do Trout
Hear?— Misapprehensions About Light— Mr.
Disraeli, P—— P—— A——, and Others—How
Things Look from Under the Water—Emotional
Illusions Leading to Misunderstandings—" Old
John's " Surprising Statement—Light Important
Only as a Symptom—Adverse Conditions—The
Ideal Morning.

IT is generally taken for granted that the
light in angling is a highly important
consideration. The assumption is reasonable. Fish differ from all other game in
respect that in relation to the sportsman
only one of their senses, that of sight,
seems to be of service.

Deer, for example, are at a much
greater advantage. Besides seeing with

their own eyes, they are quick to perceive danger through the conduct of other animals. If sheep in their neighbourhood are disturbed, the deer know that man is near, and are alert, probably bolting, in a moment. They are quick of hearing, too. If one may judge from the silence which the stalker imposes even when far off, a man's footfall a mile away may be as audible to them as it would be to an Australasian Black listening with his ear to the ground. Above all, they have a sense of smell extraordinarily acute. If you are to stalk a stag successfully, you must from the very start, which may be three miles out of range, keep to lee of him, which, as the air takes strange turns among the mountains, is no easy matter. A blunder on the part of the sportsman will enable the stag to scent danger at an incredible distance ; and then, in a double sense, the game is up.

Similarly, grouse not only see quickly : evidently they have sharp ears as well. Excepting on the few moorlands that are

still almost of primitive wildness, they are
nowadays driven towards the guns from
the opening of the season. An attempt
to shoot them over dogs would not lead
to satisfactory results. The birds would
usually rise beyond the distance within
which it is sportsmanlike to shoot.

Then, wild duck: Who that has
watched the habits of these attractive
birds, a brace of which would so conveni-
ently fill the space which at the end of the
day is often vacant in one's trout-creel,
can question that they are almost preter-
naturally equipped for the battle of life?
There are hundreds on or in the immediate
neighbourhood of the lake by the side of
which these words are written; but they
are practically as safe as they would be if
they were on the carefully guarded waters
of St. James's Park. If one lay out all
night, armed, among the reeds at the
head of the lake, where two streams run
in, a brace might be taken from the flight
of duck that are often seen there in the
morning; also, it might be possible, at

any time of day, to stalk the wild fowl in a slowly-moving boat with a screen of bushes in the bow; but the most cunning attempt to get at them in a candid way would be a failure. Wherever they may be resting, their position is always such that they are forewarned of your approach from the front, or from the rear, or on either flank.

The trout are in quite different case. They seem not to hear. At any rate, if they do hear, they are never, so far as one can judge, disturbed by noise. They show no sign of alarm when a railway train rushes over a bridge above the stream in which they are lying, or rising; often they are equally unconcerned amid the loudest peals of thunder. They must, it is true, have the sense of smell. Only on that assumption is it possible to account for their taking a worm, or a gentle, or a piece of roe, or the grub of a wasp, or that of a stone-fly, in flood water too thick to be seen through; but their sense of smell seems to be only a guidance

to their food, not a sense through which
they are warned of the approach of foes.
They never fly from a man until they see
him. For safety against their enemies,
that is to say, trout practically depend
upon their eyes alone. After they are
hooked, their strength, and the instinct
that leads a few of them to run into weeds
or other cover, may be of use ; but
their eyes are their primary and main
defence. It is reasonable to assume, then,
that their eyes are sharper than those of
most creatures.

That being so, it is not surprising that
when we go fishing we are anxious about
the light. What is wanted, it is commonly
supposed, is a light that will blot out the
rough edges of the tackle, soften down
any excess of gaudiness in the flies, and
make the lures look natural.

What is this light ? The answers by
any dozen anglers, even if they were
men of much experience, would be of
striking variety. One would say that a
dull day is the best. Perhaps that would

be the general opinion. It is noticeable
that Mr. Disraeli and other novelists who
are careful about local colour usually have
the sky well clouded when hero or heroine,
or both, set out to fish by the banks of
some romantic stream. Each of the rest
of our dozen witnesses might have a theory
of his own. As a rule it would be a nega-
tive theory. "A glare on the water"
would be the bane of one ; another would
like a thin veil of fleecy clouds ; another
would prefer the light of a day, characteristic of April, on which the sun is hidden
and peeps out alternately ; another would
have but little hope if the ripples were
tipped with silvery gleams ; another would
dread "lanes of light" lying upon the
surface of the water ; others, according to
individual fancies, would think well of any
light in which the water was not too blue,
or too gray, or too yellow, or too red, or
too green, or too purple. Probably the
only thought on which all would be
unanimous is that the light which falls
from a cloudless sky would never do at

all. It is generally supposed that good sport is not to be had in unmitigated sunshine.

At first it may seem presumptuous on the part of a single fisherman to question the opinions of all these twelve gentlemen ; but it is not really so. If all the twelve were of the same mind, the single fisherman might be considered arrogant ; but, as each of the twelve is assumed to have a theory differing from every one of the others, the criticism is merely a modest contribution of the thirteenth.

There is a general objection to almost all of the theories mentioned. It is that they are based on a strangely unscientific understanding of the nature of light. Take the lanes-of-light notion. It was first stated to me on Clatto, a lake in Fife, by P—— P—— A——, a man of exceptional intellect whose attainments in sport and in the criticism of literature are a tradition held in respect and affection at the Universities of St. Andrews and Edinburgh. "We shan't get many to-day,"

he remarked, when we had been on the
water, without a rise, for nearly an hour.
"No?" "I fear not," he said quietly:
"I never knew them rising well when
there were lanes of light." With a slow
wave of his left hand, he indicated the
offending glimmer. Was it possible that
this eminent thinker, P. P. A., actually
supposed that the light was distributed
in lanes? The surmise was disquieting,
and I ventured to remark that there was
not really any lane of light: the light was
all over the water, though only a section
of it was seen by us: the same illusion
would be always produced by the sun, or
the moon, or a solitary star, if the boat
happened to be drifting towards the
source of light: if it were drifting any
other way, there would be no visible "lane
of light" at all. Incredible as it seems,
my surmise was not unfounded. My distinguished friend had not been consciously
using a figure of speech when he noted
the "lanes of light." After a moment's
reflection, he said, "Ah! Just so. I

thought it was the local reflection of that little break in the clouds near where the sun is; but I see it must be the same all over the loch."

This surprising incident seemed to warrant a spirit of inquiry into other assumptions about the light in angling. Not long afterwards I was with another man on the same lake.   It was morning; the wind was from the east, which, as Fife is on the east coast, was not a bad portent; we had just begun our first drift.   "What do you think of the water?" I asked. "Splendid," he answered, gaily.   "Rough and blue; no glare; the very water I like to see!" "Yes? Look round, then." He turned; and saw that all the broad expanse behind was sparkling as if it had been studded with diamonds.   "What if the wind changed, and we had to cast in that direction?"   "O," said my companion ruefully, "it would never do at all: not a fish would stir!"   "Well, it's all the same where you throw the flies.   The glare's in front as well as behind.   Don't you

perceive ? " He perceived ; but the truth
had never before occurred to him.

It is only, however, a part of the truth
which those two incidents revealed. The
question to be considered is much less
how the phenomena of light impress the
angler than how they impress the fish.

From the nature of things, a complete
solution of this problem is impossible.
Even if we could lie under the water and
look upwards, we could have no assurance
that our vision of things would be identical
with that of a trout. The trout would
detect objects that escaped us, and those
which were visible to both would be seen
differently. The trout could tell a dun-
winged fly with a claret body from a dun-
winged fly with a red body ; but to the
human eye the flies would be very much
alike from three feet under the surface.

Still, there is a respect in which,
looking upwards into the air, the trout
and the human observer would be at one ;
and this unity is of great importance in
relation to the general assumption that

what the sportsman sees on the water
from above the trout sees from below.
To a fish or a man looking straight up
at noon from a stream or a lake on the
equator, there would be a glare; but it
would be the direct glare of the sun itself,
not the reflection of its light. In a water
of our own latitudes the sun would disturb the vision only when trout or man
had cause to look aslant towards some
southern quarter. The disturbance might
put the man off rising if there were something in the glare which it would be good
to snatch; but it does not seem reasonable
to suppose that it would keep down the
trout. On the contrary, it should bring
him up. Even if a trout can look at the
sun as an eagle is said to do, the extreme
dazzle of the light must surely blur the
shape and colours of a fly; and if the fish
thinks that some object between its eyes
and the sun is a desirable insect, surely he
must rush at it more rashly than he would
rise at a fly floating in a light permit-
ting of critical inspection ? However this

may be, the really important consideration is that, unless, indeed, there be sometimes a mirage athwart the clouds such as
there is occasionally in the desert, the
surface of the water, seen from below, can
never have any glare at all.   From above,
a river or a lake is a mirror, reflecting the
skies and all that in them is, as well as
upstanding objects on the shores ; but
from below it is no more a mirror than is
a sheet of glass without a backing of
silver.   Thus, none of the phenomena of
light which disturb the angler are in the
consciousness of the trout at all.   To
them, saving amid the exceptional circumstances for which we have made provisional
allowance, there is no glare, howsoever
fiercely the sun may blaze ; no lane of
light, even when their glance is eastward
at the dawn ; they never see on the surface
the blue reflection of the undimmed sky,
or the dingy-yellow of the snow-storm,
or the inky - purple of the thunder-
cloud.

Are we to conclude, then, that the

light is of no importance when one goes
fishing ?

That would be as empirical as any of
the misapprehensions I have endeavoured
to explain away.   It is rather more than
possible that there may be some truth in
a few of the accepted understandings on
the subject.   What that truth may be I
will show immediately.   For the moment
let us note how easily, on such an illusive
subject, misapprehensions, which become
convictions, arise.

Only a few of us have the good fortune
to fish continuously for months.   The
rest have to be content with a day, or a
few days, at a time.   In most cases, then,
our craft in angling is derived from experiences far from complete.   Nevertheless, it is a settled body of doctrine, of
principles unshakably fixed.   Our observations by the riverside, or on the lake, are
vivid and memorable for their rarity.
We had a week, let us say, at Whitsuntide, and sport was good on all the days
but one.   What are our recollections ?   A

little introspection will lead to an illuminating discovery. The recollections are in two classes, one of which is vivid in general joy, while the other is vivid in detailed distress. Of the good days we remember how cheerfully the trout rose, where we landed the three-pounder, where the bigger fish broke off, and what merry nights began when we all assembled at dinner; but whence the wind on these days? Did the sun shine brightly, or was the sky clouded? Were the days warm, or were they chill? Was the weather fair, or were there showers of rain? Our recollections on these points, it will be found are vague. The sport and the mirthful happiness are very fresh memories indeed; but all we can say about the weather is that, whatever the details may have been, it was certainly exceeding good. Then, the day when the sport was poor: Ah, there is no difficulty on that score! The morning was promising enough; but we had not been out for an hour before we discovered that the wind was shifty.

We remember the very moment when it came sleet-strewn from the north. It died down while we were seated at luncheon under that old oak on the meadow near the farmhouse. Then the light clouds slowly thickened until the whole sky was slaty-gray; and about seven o'clock, just when the evening rise should have come on, the sun flared out angrily among storm - clouds scarlet and green and yellow. All the time scarcely a trout would rise. Now, not one principle of angling, but a whole series of principles, naturally springs from the observations of an unfortunate day such as that. The series is, That trout do not rise when the wind is shifty; that the northerly breeze, especially with sleet on its wings, is bad; that a languid afternoon following a fresh morning is worse; and that sport is altogether out of the question when the heavens at sundown are on fire. The consequence is that when one falls upon such a day again one either puts the rod into its case or uses it in the perfunctory manner of the hopeless. One does not expect sport, and does not offer the fish a fair chance to give it.

Such is the genesis of almost all our principles of angling, which, it will be observed, are principles of taboo. It is much easier for any of us to say what weather will not do than it is to say what will; but are we generally right in our taboos? I doubt it; and, as I have made careful experiments, there is cause for the cheerful misgiving. One May afternoon I fished carefully over three miles of well-stocked water, and returned with an empty creel. There was a little wind from the west, sufficient to make an attractive ripple here and there; but how languid the gray clouds were, and the air how life-less! Suddenly, and without premeditation, I said, "Is it really so? Would the sky and the air seem languorous and dull if I had filled my basket to the brim, as a few days ago I filled it on this very stretch?" Truthfully I could answer that they would not. The grayness and the languor

were just as much subjective as objective.
They may have been on the sky and in
the air; but they were also, and I think
primarily, within, affecting the outlook.
Certain it is that my recollections of that
day's weather, which, after all, was normal
for the time of the year, would, though
general, have been wholly favourable had
sport been good. Often the gloomy
aspect of the weather is only an emotional
illusion.

If, then, we would be really skilled in
the craft of angling, it is necessary that
we should be much more careful in our
deductions and our inductions than most
of us habitually are. These processes of
reasoning are apt to become entangled to
our confusion. It has been admitted that
there may be some truth in the beliefs
that much sport is not to be expected
when the water is flagrant in the sunshine;
but this admission is not by any means
absolute. The beliefs call for explanatory
interpretation, which may best be given
by stating them in a new way. It is not

the "lane of light" in itself, not in itself
the glare on the water, that keeps the
trout down, when it does have that effect :
it is some atmospherical condition of
which the "lane of light" or the glare is
a symptom or a casual incident.   That
conclusion is forced upon us by considerations which no observant angler can call
in question.   Apart from times when the
whole sky is overcast with a heavy and
unbroken cloud, there is not a single day
in the year when, if we looked upon the
water in the direction of the sun, we
should find to be missing all the objectionable phenomena of light.   One of them,
or some of them, or all of them, would be
before our astonished eyes.   It follows
that if the phenomena were really as
objectionable as they are supposed to be
there could never be a good day's sport
at all.   As there is many a good day in
the season, it is clear that the taboos are
unwarranted.

It may be that the lane of light or the
glare has been witnessed on a day, or on

days, of disappointment in the pursuit of
fish ; indeed, having regard to the general
belief that the streak and the glare are
unfavourable, one easily perceives that
it must have been ; but what does this
prove ? It does not prove very much.
Those who have a day on the water only
at rare intervals take it for granted that a
good one is just as likely as a bad one to
fall to their luck ; but that is a mistake.
After a rather dull outing on Lochleven,
I remarked to Old John, the boatman,
that, although I had fished there two or
three times a season for five years, I had
never yet chanced upon a really good day.
" I can believe that," answered the vener-
able man. " I mysel' ha'e been fishin'
this loch for sixty years, an' I've seen
only one really good day." That was a
startling account of a water which is
famous among sportsmen all over the
world ; but, howsoever exacting Old
John's estimate of a good day may have
been, there was more than a grain of
truth in it. A good day is not the rule.

It is the exception. This will be found
out by any one who fishes every day for
a month. As I write these words I am
in the midst of an even ampler experience.
On most days during the latter half of
March and the beginning of April sport
was good; after that, for nearly a month,
it was on most days poor; since then, on
a few days, there have been signs of a
revival. Is not the moral manifest?
The chances are that if I had been
on the water only one day, instead of for
many days consecutively, it would have
been a day of poor results; and probably
that would have been attributed, con-
scientiously but without much thought,
to the aspect of the weather, in which,
as a rule, the quality of the light is the
most noticeable phenomenon.

If the other conditions of the atmosphere were taken into account, it would
soon be surmised that the light is not as
a rule the cause of either good sport or
bad sport. It may be a symptom of the
cause; but in itself it is only incidental.

Light being of many varieties in intensity and in colour, a whole volume would be required for a discussion of it that would even approach completeness ; but there are two indisputable facts touching our present theme. One of them is as yet a complete perplexity. The other, I think, will be acknowledged as evidence that most of the taboos we have been considering are superstitious.

The first fact was pointed out to me by a gillie in the Highlands. " They'll be dour thi' day, I doobt," he said, as we launched the boat one morning in the spring. " I never knew them takin' when thae misty clouds are sittin' on the hills." Sure enough, that day the trout were dour indeed. Only one, and that small, was the reward of a long and assiduous effort. This was remarkable. The soft wind was pleasant, the light was all that any angler sensitive on that point could wish, the mercury in the weather-glass stood at " Fair "; yet the trout would not rise until the clouds floated upwards from the hills, or were dissolved. In spring and the beginning of summer, mornings such as that are frequent in the Highlands; and, observing carefully, I have never known the gillie's rule to fail. In England, too, I have often had testimony to its truth. There the symptom of the peculiar weather is not so easily discoverable; but usually, even in Hampshire, which is comparatively flat, there is not far away from the water some ground that rises high enough to show it; and in England, as in Scotland, the trout keep down when the misty clouds hang low. The explanation is beyond me; but most anglers, I think, will agree that it cannot lie in the quality of the light.

The second of our indisputable facts is much more pleasant to contemplate. Who does not recall many a morning on which the fish, in lake or stream, rose well while the blue water, under the south-west breeze, twinkled in the unclouded sunshine? Usually on such days I myself, at least, expect good sport; and nearly

always on such days I find it.   The light
is as brilliant as it can be ; yet the fish
are not made shy.   Surely, then, the
belief that a strong light keeps them
down must be abandoned.   A belief that
it brings them up, which impetuous
reasoning might suggest, would be equally
untenable.   On a day such as we have
been considering sport is good simply
because the conditions of the weather, of
which the light is only a single symptom,
are all of them favourable.

What these conditions exactly are it
would be rash to say ; but I have noticed
that they are always present during the
period between the passing of a cyclonic
storm-centre and the complete establish-
ment of a high-pressure system of varying
light breezes or dead calm.   Sometimes
the trout feed while the storm is rising,
and sometimes even when it is altogether
past ; but sometimes they do not.   The
only time when I feel absolutely certain
of good sport is when the barometer is
rising in the recovery of the atmosphere

from an outbreak of lightning and the wind. When the recovery is complete the sport becomes inconstant. Then, howsoever agreeable the weather may be to society at large, to the angler it is a speculative risk. The trout may rise freely; but that they may not is just as probable. Indeed, it is more probable. There are a few half-public waters the sport on which is regularly reported in the newspapers throughout the season. If one watches the tidings, it will be found that for every really good day there are at least twenty indifferent or bad days. This astonishing fact, which will be considered in another chapter, and there shown to be auspicious, means, among other things, that the climate of the British Islands is much stabler than it is commonly reputed. There are many small changes in the weather; but great changes, storms, are infrequent.

CHAPTER V

ARE TROUT CUNNING ?

England and Scotland—The Scotsman's Better Fortune and Less Keen Interest—Scorn for " Fancy Flies "—Midges Everywhere—Mr. John Gilbert's Wonderful Basket—His Large Flies—"No Rise," Yet Good Sport — Trout have Unmistakable Preferences—The Fish are Not Capricious—Do they Become Wary ?—T—— J—— B—— and his Chalk-stream—The Adventure of Mr. T—— — The Forbidden but Instructive Otter — " Wariness " Apparently an Illusion—Suggestions towards Accurate Knowledge—A Memorable Morning.

FLIES are better understood in England than they are in Scotland. That, perhaps, is mainly because opportunities to use them are much more frequent in Scotland than in England. Even as many thousands of Londoners are unfamiliar with historical buildings in the Capital, dwellers in regions where fishing is to be had for the taking or the

110

asking, or at small charge, have but a casual interest in the sport.

In England trout-streams are rare, and trout lakes rarer; and the waters are in most cases private. In England a day's fishing is either a costly luxury or a great privilege. In the South it is no uncommon thing for a club of twenty-five men to pay £1250 yearly for the right of fishing in two or three miles of stream. Considerations of that kind stimulate the imagination, and English anglers set themselves to become as proficient as possible in the craft of the sport. They may still be far short of the complete science or the perfect art; but they try to be expert in both. In Scotland quite a different attitude is the rule. Almost any one there can have a day's fishing, or a week's, if he wishes to, and has time to spare; but he does not make the best possible use of his privilege. He seems to regard angling as an amusement in which to pass the time pleasantly, rather than as a craft to be closely studied.

Indeed, there is some cause for suspecting that the people of Scotland do not
really believe that there is, or can be, a
science of the sport at all.   They are disposed to smile when any poor Southerner
appears among them equipped with all the
tackle which a first-rate shop in London
can supply.   Nearly all of it is superfluous,
they think ; and the rest is probably
shoddy stuff.   All that's wanted, they
will add when frank in their friendliness,
is to be had at the local ironmonger's.   In
fact, excepting in Edinburgh, Perth,
Inverness, Glasgow, and one or two other
towns, the ironmonger is the recognised
authority.   Often, as he spreads out his
cases, he will show you a really wide
variety of flies ; but of most of them he
has a poor opinion.   In spirit, as now and
then in act when there is nothing to do,
he is a sportsman ; and he does not conceal his opinion out of consideration for
his trade.   His opinion echoes the voice
of the people who go to his shop for flies,
and "bait hooks" when there is a spate,

and new lines when the old ones are
wearing a little rotten; and usually it is
remarkable for simplicity and emphasis.
"Nae use ava'," he will often say, in
cheerful condemnation of a whole boxful
of what he calls "fancy flees": "here are
the flees that tak' a' the year roond,"
opening a case which contains blae-wings
and woodcocks, some with red hackles,
others with black, others with buskings of
hare's-ear, and a few flies of hackle only.

If your visit to the good man is before
the end of April, he will commend to you
pretty large flies; but after that midges
only have his approval. Morning, or
noon, or night, it is only midges, he
assures you, that are any use. "Even
doon the burn, where the tide comes up
frae the sea, the water," he explains, "is
sair hard-fished nooadays, and the troot
are awfu' cunnin', and wi'na' look at onything but midges." Here and there you
may find a trader in tackle who says something else; but all the tales are variants
of the same rather pessimistic unbelief

in the "fancy notions of tourists frae Lunnon." Blae-wing, woodcock, and the hackles, small for streams, larger for lochs, and for lochs supplemented by teals and perhaps a few heckham-peckhams, will serve all over Scotland.

So the local authorities say, and the local anglers achieve wonderful results with the limited equipment. Living constantly within easy reach of streams and lakes, Scotsmen are nearly all of them anglers more or less, and those who fish frequently fish well; but they do not realise that the craft has great possibilities of refining and development. Indeed, in sports and pastimes generally they seem to be constitutionally content with mediocrity. In football they are, I think, supreme; but there is no other game of skill, no other sport, in which they are equal to the English or to the Irish. This is notable in regard to cricket, in which the best team raised from the whole of Scotland would probably be no match for a second-class English county; and still more strangely notable in regard to golf, the very implements of which remained rather rude and ill-adapted to their purposes until, after centuries of perfect contentment with them among the Northmen, the game suddenly spread into England.

Angling is not exactly analagous. Although there is less scope for it in the South than in the North, angling is in England a sport as ancient as it is in Scotland. Nevertheless, we see as regards angling the difference between Scotland and England that marks their standings in other recreative pursuits. The Englishmen are keen and progressive. The Scots are indifferent and stationary. Nay : it may be said that they are retrograde. Signs are not wanting that two or three generations ago the contents of their tackle - books, or at least the flies they actually used, were more reasonable than those of to-day.

One fine afternoon in mid-summer I came upon an old gentleman preparing to

fish in a broad rough pool under a waterfall on the Fife Eden. He had not, he told
me, been on the river for very many years;
but the weather had been pleasant that
morning, and he had thought of coming
out to cast a fly. Wishing him good luck,
I passed on ; and, having fished diligently
for two hours not far off, I wandered back
towards home, and came upon the old
gentleman where I had left him. My own
basket held a brace of trout, each fish
about half a pound. I wondered, Had
even so much as this modest fortune
come the way of the ancient sportsman ?

Lying beside him on the grassy bank
where he was seated were the finest five
trout I had ever known to be taken from
that stream or any other ! Each of them
seemed to be well over two pounds.

As I was gazing upon them, almost
doubting the evidence of my eyes, the old
gentleman said, "Hullo ! They're really
on the feed to-day"; and, looking up,
I saw, from the bend of his rod, that
he had hooked another. He played it,

and landed it; and it matched the five
fish well, fat and firm and shapely, white
of belly, and with bright yellow sides
spangled by clear-red and dark-blue spots,
a trout in prime condition.

How did he do it? Was this nona-
genarian a wizard? or, by birth an Irish-
man, bearing the honoured name of
Gilbert, had he the secret of those strange
essences with which his volatile country-
men in remote parts, where water-horses
were still not unknown, while gnomes
and sprites that haunt the nights were
common, smeared their lures, to the complete undoing of the fish?

Apparently, it turned out, there had
been no black art in the old gentleman's
triumph. In response to my awe-struck
questionings, he showed me his flies. In
obedience to the ironmonger, I myself
had been using midges. The old gentleman's flies were three or four sizes larger:
indeed, they seemed just the flies that
one would cast upon Lochleven in a
breeze!

" They were in me ould book," he said,
quite simply, though very happy. " I
had them when I came to these parts
with Mr. Erskine Wemyss, contesting
the county at the time of the Reform
Bill. I thought they might be a bit
weak by this time ; but they're none so
bad, after all."

This was suggestive ; and a very little
thought led to one's seeing things which,
although they had been daily before the
very eyes of all frequenters of the stream,
had been completely unnoticed. It was
not only midges that were about. All
over the water large flies of many hues,
with here and there a buzzing column
of alders, were fluttering. The old
gentleman's day on the river was full
of marvellous sport simply because,
using large flies, he had followed
an old tradition suggested by Nature,
instead of following the precepts of the
ironmonger.

It may have chanced that, besides
being of the right size, his flies were of

the right colours and the right shapes.
As the old gentleman had been so successful that he could hardly have been
more so, it may be said that obviously
they must have been right in all respects.
This raises a delicate problem.

Should our flies invariably be imitations
of those which are on the water at the
time of fishing?

English fishermen, especially those
who frequent the southern chalk-streams,
where angling is a high art, think so, and
I myself, in a general manner, share their
opinion; but there are great dubieties to
be resolved before absolute rules can be
formulated. Early in the season one
frequently catches trout after trout, quite
quickly, when neither a natural fly nor
a "natural rise" is to be seen. How
is that? How can it be said that our
lures are correct imitations when there
are no flies to imitate? The absence of
natural flies is not quite exceptional. It
is frequent. On what principle, then, do
we accept certain lures as appropriate

to times when the creatures they imitate
are dead or still unborn ?

As suggested in an earlier chapter, we
accept them on the understanding that,
although the real flies are not to be seen
on the day of our fishing, they would be
visible if the weather were more propitious.   Resembling a famous golfer, Spring
never plays up to her average ; but there
really is an average if only we take into
account a sufficiency of years.   For example, among the flies for the opening
month of the season are the March
Brown, the Woodcock and Hare's-Ear,
the February Red, and the Black and
Blae.   It may be that some fisherman
of long ago found on the water flies
which those lures resembled, and there-
upon established a tradition which has
come down from age to age.

An alternative theory is that at the
beginning of spring, when flies are very
scarce, the trout, being hungry, rise at
anything that seems to be a fly at all,
without curiosity as to which of the

possible insects it may be. This thought
would find general approval in Scotland,
where even good fishermen say that if the
trout are really on the feed they will rise
at anything, and in Yorkshire, where
fishermen, when out on the becks, have
only a few flies, all hackles, which are
deemed sufficient at any time of the
season ; but it will not commend itself to
any one who has been closely observant
in angling from day to day.

Even on a little-frequented mountain-
burn or moorland-stream, on which artificial lures are thrown only once or twice
a year, the trout invariably show a marked
preference for some particular fly. I have
never known an exception to this rule.
It is true that the trout in burns and
becks rise freely, and that you are almost
certain to catch some with whatever lures,
if they be not of unreasonable size, are on
the cast ; but this only makes their preference the more remarkable. Sometimes
it is a fly with a red body that attracts
them, sometimes a fly with a brown body,

sometimes a fly with a black body, and sometimes a fly without wings. You may have to try a good many lures before you hit upon the right one; but when you do there is no room for doubt. For every fish that takes either of the other flies two or three take the right one. Often, too, the trout change their preference from day to day.

What happens on unfrequented waters, to which I have referred because there the trout are most obviously in a state of nature, happens on streams and lakes that are whipped the whole season through.

When the natural flies by which we could interpret the preferences are, as often happens, not to be seen, all this is very puzzling; but it cannot be attributed to caprice. Far from being capricious, trout, I seriously think, are not even capable of acquiring wariness. Often we hear that the fish in such-and-such a river are very cunning. "It is so much fished," we are assured, "that it takes the very highest skill to catch them. Not like the good old days." There can be no doubt that in saying this one's friend has a definite thought which he states in all sincerity ; but if we reflect a little we shall find cause for suspecting that he is in every case mistaken.

The assumption that trout in much-fished waters have become wary is based upon the fact that many of them, having been hooked and lost, lived to fight on other days with more than their native discretion. The statement of fact may be conceded to the full, even to the extent of admitting that in certain streams every trout old enough to rise at flies has conceivably at some time or another been pricked by a hook ; but the inference is exceedingly doubtful.

One is loath to tell stories which almost anyone of experience could cap ; but it is just because an incident which I now recall could easily be matched from the recollections of many another angler that I set it down. The certainty that it will bring to memory similar incidents in the experience of others will go far towards rendering credible my conjecture that trout are not now more "sophisticated" than they ever were.

That blithe sportsman by flood and field, Mr. T—— J—— B——, possesses a three-mile stretch of a stream rising on the borders of Surrey, Sussex, and Hampshire. Year by year it is his hospitable custom, when, in the pressure of business affairs, he remembers that it is time for fishing, to invite friends to go with him for a few days to H——. As the thoroughly ancient Royal —— Hotel has accommodation enough for six guests besides himself, B—— is rarely without half a dozen boon companions when he entrains at Waterloo. It does not matter at all that one or two of them may never have waved a rod before : at least, if this is taken into consideration, it does but lend additional gaiety to B——'s view of the outing. He rejoices in the joy of a friend over the exciting marvel of the first

trout. In fact, the water is peculiarly
adapted to B——'s versatile hospitality.
Besides the stream properly so called,
there are two ponds, embanked large
pools, fed by the stream itself, from out of
which go the forces that drive the wheels
of a quaint old mill by the wayside ; and
in these ponds are many trout of enormous
size, some of them, which are seen now
and then, believed to be twelve pounds
each. The ponds are what may be called
B——'s reserves against the possibility,
not to be tolerated for a single convivial
moment, that some one of his guests
might have the hilarity of dinner blighted
by the remembrance of an empty creel.
If any of his party looks like being
defeated on the stream, B—— knows
what to do.

Well, on our setting out after breakfast one morning, he told me that I was
to keep an eye on Mr. T——. Mr. T——
is well advanced in years. His high
and esteemed position in the City of
London is the result of such a busy

life that he had never had any time to
fish, and that was his first day with the
rod. If he did not have a trout pretty
early in the afternoon, I was to lead him
to the ponds, and see that he got one
there. "And, mind you," B—— added,
"it is to be a good fish—big enough to
make Mr. T—— want to have it set up,
to be an heirloom in his family for
ever."

In due time, according to these instructions, I led Mr. T—— to the ponds ; and,
arrived there, asked to see his flies.   The
end one had a peculiar dark wing and a
body of claret colour : once seen, it could
be easily recognised on occasion.   The
hooks were somewhat large for the purpose which B—— had commanded me to
see accomplished ; but they might as well,
I thought, be given a trial.   If they failed,
a fresh set could be readily fitted up.   On
each of the three I put a worm ; then I
cast well out towards the middle of the
pond, handed the rod to Mr. T——, and
pleaded with him to raise it smartly, but

without violence, when I should say
"Now!"

In less than a minute the word of
anxious command had to be given ; and
in less than another my new friend was
pleading with me. "Take it—take it!" he
exclaimed, trying to hand me the rod,
which, in much perturbation, was bend-
ing and wriggling in all directions. I
took it, and spoke soothing words ; and
when Mr. T—— had recovered a little
from the first shock, in duty bound to our
host I cajoled him into risking another.
"Keep up the point of the rod," I said as
he began again, "and let the line run out
when he pulls hard ; and all will yet be
well."

For a few minutes, though he was
visibly trembling and only gasped when
he wished to speak, Mr. T—— managed
all right ; but then a pair of peasants came
along the road bordering the ponds, and
stopped to look on, and apparently caused
Mr. T—— to be forgetful. At any rate,
when the fish bolted again, he allowed the

rod to be pulled down till the point
touched the water, and clung to his gear
with might and main. The line suddenly
slackened. The trout was off. So, I
found on examining the cast, was the
end fly

The incident, however, was not yet
closed.

I put another cast on Mr. T——'s line,
and, after waiting for ten minutes in order
that the many fish in the pool might
recover from the disturbance, persuaded
him to try again. The bait was seized
soon after it fell upon the water; there
ensued a dire struggle, which lasted nigh
half an hour; and in the end, all a-tremble
and full of laughter, scarcely able to keep
from gambolling in his glee, our friend
was in proud possession of the heirloom
which B—— had desired for him.

In the mouth of the trout was the
lost fly !

The fish Mr. T—— had caught was
the very trout he had been struggling
with, and had lost, when the peasants

came upon the scene. It weighed four
pounds and a half. Surely, therefore, it
was old enough to have learned wariness
by experience if trout are capable of such
learning at all.

There are, as I have said, strong
reasons for suspecting that they do not
have this capacity. Incidents carrying
the same suggestion which comes irresist-
ibly from Mr. T——'s performance are
not uncommon. A trout often rises at
the very fly which a few seconds before
severely jagged him. Most of us know
that he will sometimes go on rising and
being jagged again and again, just as if he
were determined to be caught. Not all
the trout one meets behave in this way ;
but many of them do, and their conduct
casts doubt upon the belief that the fish
are taught by experience to shun artificial
flies. Not only do they seem to learn
nothing from their own mishaps : they
seem to learn nothing from one another.
Often when you are playing a trout, a
second seizes another of your flies. As

9

the trouble of the first fish may conceivably be not manifest to the other, or not interpretable by him, this counts for little ; but what are we to make of the otter ?

The otter is a gear, now forbidden by the law, taking a line of gut from which depend two or three dozen flies, each about a foot apart from its neighbours, far out on a river or on a lake. As the poacher, with one end of the line in hand, moves along the shore, the otter-board, which is constructed on the principle of the kite, moves onward too, and outward ; just as the kite follows the holiday-making school-boy, and soars at the pressure of the wind. Soon the poacher feels a tugging at the line, and knows that one of the flies has been taken by a trout ; but he does not reel up. On he moves, and soon there is a fresh tugging, and soon, he thinks, another ; but after that he is not sure how his adventure fares. There may be new tuggings ; but they cannot be distinguished amid the old ones. He can tell nothing, either, by the weight of the line : besides

keeping the line taut by a mechanical
necessity, which only a salmon or a very
large trout could undo, the otter-board is
heavily weighted with lead on the sunken
rim, just as the kite is weighted on the
tail : by the sense of touch through his
rod and line, the poacher would know no
difference between three ordinary trout on
the hooks and three dozen. All he can
say to himself is that if he goes on a little
longer each of the flies will in all probability be taken by a trout; and he is right.
Soon he turns, and goes back upon his
tracks ; and the otter-board comes in, just
as the kite would fall if the schoolboy
could cause the knot of the flying-string
to slip down past the middle of the belly-
band ; and lo ! as the poacher's line comes
in a trout is dangling from every hook.

More than one moral for the instruction of the angler could be drawn. A
thought that is instant and insistent is
that the success of the otter is complete
disproof of the theory that the dry fly
is on waters which are much fished the

best lure for trout, if, indeed, it be not
the only one in which there is any
hope. All the poacher's flies have been
at least a foot below the surface; yet
the trout have found them irresistible.

That, however, is not the thought
which is at present relevant. I have
described the working of the otter in
order to show that the capture of its
own kind bears no warning to the trout.
All the two or three dozen fish have
impaled themselves on the same line.

Those who have been in the habit of
assuming that trout acquire wariness may
endeavour to explain this away by saying
that the trout did not know the things
depending from the poacher's line to be
artificial flies. That would be to yield the
whole ground on which their assumption
rests. It is only on the assumption that
a trout often knows an artificial fly when
he sees it that wariness in relation to
artificial flies can be attributed to the fish.
If the fish lack this knowledge, as the
success of the otter joins with the ex-

perience of the sportsman to suggest, it is clear that adversity does not teach them, and that the wariness with which they are credited is an illusion.

Beyond a doubt, it is. Reasoning will assure us so. Some streams yield less good baskets than they yielded in days gone by; but, if the subject is carefully looked into, it will be discovered that the falling-off may be explained as referable to a thinning-out in the stock of fish. Anglers increase in number year by year. It is not surprising that on waters which are said to have yielded three dozen trout in a day fifty years ago three or four brace in a day are now considered a fair basket. This is particularly applicable to Scotland, where there are many streams open to the public. The anglers now are at least ten times as many as they were in the days of their grandfathers; the "free waters" are never re-stocked, or hardly ever; it is only in the nature of things that as the sportsmen have been gradually multiplying the head of fish has been gradually dwindling. Elsewhere the balance between the state of affairs in the olden time and that of to-day has been kept nearly as it was. It has been so on all the private lakes and streams in Scotland and in England. What, then, do we find there ?

The sport being private, statistics are not accessible ; but the question does not lie in impenetrable darkness. Memory plays us false by certain illusions which are corrigible on calm reflection. From time immemorial we have been talking about a change in the climate. Winters, we feel, are much less severe than they were wont to be, and the summers much less fine. Official records show that since the Meteorological Offices were established there has been no change in the climate to justify that generalisation. The truth is that, especially if we rejoice, as is proper, in pastimes on the ice, "good old-fashioned winters" linger in our memories, while the mild ones which intervened fade away, and are forgotten ;

and the past seems, in one half of it, a
succession of winters when the sound of
skates and curling-stones was continuous
in the resonant air, while the heavily-
laden postman was always knee-deep in
Christmas snows, and, in the other half, a
succession of summers so glorious that
only the thought of partridges and
pheasants and grouse and red-deer when
the russet leaves should begin to fall, and
of the fox hunt when the trees were bare,
reconciled us to the passing of the sunshine. In these things memory knows no
breach of continuity.

It plays us the same trick as regards
sport on the streams and lakes. All the
good days of ten years ago are vividly
remembered, and all the poor ones are
forgotten. In the present it is the poor
days that are conspicuous : the good ones
will not reach their ultimate grandeur
until another ten years have elapsed, and
then they will appear to us as a season
of unbroken brilliance. Can any one
honestly say of a well-preserved water

which he has given fair trial, for not less
than the customary number of holidays,
that this season his sport has been less
good than usual ?   Then, if it has been as
good as usual even on a single day, is
that not clear proof that it may yet be
as good as ever on many days ? Had
the trout really become as wary as many
of us suppose, they would not rise as
freely as of yore even for a single day.

It may be said, If the trout have
learned nothing by experience, and, as the
poacher with his otter seems to show,
cannot discover the ruse which lies in a
fly of steel and fur and feather, why do
not they always rise readily at our lures ?

Some of the reasons have been indicated in earlier chapters.   They resolve
themselves into the knowledge that there
are conditions of the weather in which
sport is dull, and that the conditions, or
some of them, are very common.   The
trout are not intelligently capricious.
Only, they are as sensitive to the atmospherical conditions as the barometer it-

self. That is their natural safeguard.
If they always felt an impulse towards
our lures, soon there would be very few
of them left. Nature preserves their
species by providing on most days of the
season an atmospherical sanctuary.

Still, I am confident that many a bad
day would be not quite so bad if only we
had a wider and more accurate knowledge
of flies. Many a time does one fish for
two or three hours before, after changes
of casts, the proper fly is brought into
play. Then the sport is frequently so
good that one is tempted to think that
there would never be a blank day if only
one were nearer complete knowledge of
the flies. If the right fly for the day
were always on the water punctually,
Nature herself would give us the broadest
of hints; but, especially on lakes, Nature
is frequently taciturn. It is not unreasonable to think that there is a right fly for
every day; but one has usually to find
it for oneself. This is not altogether to
be deplored. In sport, as in philosophy,

the pursuit of the truth is fascinating;
and in sport, whilst not always in philosophy, discovery of the truth is an unmitigated delight.

That being so, a disclosure of secrets
wrested from Nature on streams and lakes
might be considered not an unqualified
boon; yet I am not without defence in
deeming it unnecessary to withhold The
Book of Flies which is inset at the beginning of this volume. While proving, I
shall trust, helpful in no inconsiderable
measure, it will not lead to depletion of
the lakes and streams. Thousands of
anglers might become much more expert
than they are without making any perceptible reduction in the stock of fish
those waters hold.

With a defter eloquence than could
be spun from words, the Book sets forth
one's own experience and a very large
mass of carefully sifted traditions; but
an explanatory word may here be added
to what is said in the Preface. While
the pictures of the lake flies are in most

cases larger than real insects, those of the
stream flies are as nearly as possible
of the same size as the insects which they
represent. Why should we have large
flies for rough water and small flies for
water that is calm? The insects of
nature do not grow larger with the rising
of the wind and smaller as the storm
abates. They are of the same size in all
weathers. It is surprising how readily,
when they are disposed to feed, trout
see and seize a lure that one might think
lost in the tumult of windswept waters.
It is even more surprising how readily,
when in the same mood, the fish will rise
at a fly so large and gaudy that it seems
out of all proportion to a lake in calm
or to clear shallows in a stream. In
saying this I am, of course, assuming that
in both cases the fly is of natural size.

Sometimes, I think, a trout may be
induced to take a fly by being offered it
over and over again. Well do I remember the incident which first suggested
this. It fell on the day of the first trout

of the season a few years ago. He rose
to my companion. She had been casting
over a pool sheltered by banks so high
that the wind did not stir the surface in
the least. The peacefulness of the scene
was suddenly dispelled. The cast fell
close to the northern bank, where the
water was deep; the fish threw himself
upon the tail fly with a ferocity strangely
out of keeping with the placidity of things.
I will not describe the subsequent pro-
cedure in detail. There is an absurd recollection of oneself upside-down, hanging
from one's toes upon the bank and head
towards the depths, striving to reach the
trout with a short landing-net. Suffice it
to say that we did land him, and that he
was a pounder in apparently excellent
condition. In England as well as in
Scotland trout often seem to be more
forward in April than they are sure to be
in June. That has puzzled me many a
time. The explanation cannot lie in
one's natural willingness to find the first
trout of the year surprisingly hale and

hearty. Although not so plump as it
would be in June, the April fish is sometimes brighter of aspect than the summer
fish; and it is mainly by the colour that
anglers judge of a trout's condition. If
the hues are iridescent and there is no
white fringe on any fin, they say the
trout is perfect.

Well, just such a trout was the one
which the lady caught that morning. Of
course, I could not believe that it was
really a better fish than the three-pounder
which I hoped to see her playing in the
second or third day of the Mayfly week.
Perhaps the fine appearance which trout
often present early in spring is due to
the state of the stream. Besides being
cold, the water then is full and fresh.
Spring is the most vitalising of the
seasons. At any rate, in the stimulating
sunlit coolness of noon the lady's expres-
sion, as I lifted the trout from the water
to the grass, was even rosier than it
would be when she graced the ballrooms
of midsummer.

Is it only a poet's license to assume that all sentient beings may be subject to the same stimulating influence? No: it is to the weather of spring in England that English woodlands and English meadows and English maids owe their unrivalled beauty. If we protected ourselves and our country from the trying temperatures of its winter solstice, by enclosing these islands in a huge glass case which any trans-Atlantic inventor could easily construct, our women would be no more comely than those of St. Petersburg, who live in hot-houses six months of the year, and our meadows would be as uninspiring as the prairies.

It has not yet been said where we fished that day. The truth is that the name of the stream is unknown to me. In that respect one may liken oneself to a certain dandy well known in Mayfair. Shortly after joining the Fifth Hussars he met a friend belonging to his old regiment, who asked him what regiment he had joined. "O, my dear fellow," said the dandy, "I don't know its name. You go to it from Waterloo." Similarly, we went to the stream from Waterloo. It flows, westward, through a valley in Hampshire. The weather was excellent. As we were putting up the rods, a slaty-blue cloud, tinged in its lower surface as if with smoke, sat high and motionless in the north-west. Soon a wind roared over the hills, and we were pelted for ten minutes with snow of a strange dryness. That was not altogether a bad omen. When the cloud had spent its fury it would leave the valley swathed in stillness and sunshine.

It did; but, unfortunately, there was no hatch of any fly. We did not see a single "natural rise" all day; yet we caught here and there a trout. That is a matter for wonder; but it is not inexplicable. The lady, I noticed, was very slow in her movement up the stream. She was not content with one cast over any likely place. Cast after cast, to the number of at least a dozen, she made before she stepped a few yards farther. On any reasoning according to the orthodoxy of angling, one would not have expected her to find much sport on that system. One would have thought that if a trout did not rise at the first cast over him, he would be put down, and would stay down. That idea, it turned out, would have been a mistake. Half an hour after the capture of the first fish, the lady was battling with another. After having had the flies thrown over him at least twelve times, he leapt at Mellursh's Fancy with precision and honest intent. What the intent was, whether it came from appetite or from anger, one cannot say for certain; but my own interpretation of the lady's success was simple. It was that the trout had not been alarmed by her first cast, or by her second, or by any other. He had not rushed at the lure in anger. He had simply said to himself, as cast after cast dropped over him, "Upon my word, here's a rise of Mellursh! It is pretty early in the year, and Mellursh cannot be very succulent just yet; but I may as well taste and see."

Some theory of more learned aspect might have arisen were it not that when the question occurred it was lost in a more interesting consideration. The lady, who had been gathering primroses, presented to me a buttonhole. That brought to one's memory the well-known lines of Mr. Wordsworth about a primrose by the river's brim. Before then I had thought of Mr. Wordsworth as being a great poet but otherwise a melancholy wildfowl. Suddenly he was enshrined in one's regard as a great poet only.

CHAPTER VI

OLD JOHN, TIM THE TERRIER, AND
OTHERS

WHEN Old John said that in sixty years
there had been only one really good day
on Lochleven, he was not making a merely
picturesque remark.   He may have been

conscious that there was humour in his
words ; but perhaps he was conscious also
that the humour lay in a truth which they
contained. At any rate, he should have
been. A day is "really good" when it
bears comparison with the best ; and what
was that ? Was it not a day when from
morning until the night trout rose at the
flies so well that the basket was limited
only by the frequent need to spend five
or six minutes, sometimes more, in playing one, with now and then an interval
for the setting right of a tangled cast ?
There is, and always may be, such a day
as that. Happily, it is not so infrequent
on other lakes as Old John found it on
Lochleven ; and it must be known to
most anglers whose good fortune has en-
abled them to give lake-fishing a fair trial.
Why, having come once, does it not often
repeat itself ?

The answer, I think, will be suggested
if, in the chapters on Wind, Light, and
Temperature, I have succeeded in the
endeavour to show that the atmospherical

conditions of sport are exceedingly complex, and only on very rare occasions arranged exactly in favour of the angler. The angler is not the sole creature whom Nature has to consider. If there were always wind to make a curl on the water, the flowers and the fruits of the fields would be robbed of some of the warmth which the sun offers to their needs; if the clouds never settled and thickened until the whole earth seemed shrouded in a gray pall, depressing to trout and fisherman alike, there could be no rain, and vegetation would be impossible; "snowbrew" floods are inevitable, because a mantle of snow is the means by which the winter frosts are prevented from penetrating too far into the soil. When one comes to think of them, all the phenomena of our sport are clearly in the woof of a system which seems to be of design. If it were not that the trout are put beyond our reach by falling aloof at the touch of natural conditions which are very common, one of two great misfortunes would come to pass. Either, being without what at present in relation to the fisherman is an accidental instinct of self-preservation, they would speedily be all caught, leaving the waters empty of their species ; or they would be so easily caught that we should cease to think them worth pursuing, and so lose one of the greatest pleasures in life, an outdoor sport.

This may seem a crude thought, a frail suggestion on which to recall the argument about design in nature ; and I know quite well what many thinkers who may read these lines will say of it. " What ! " they will exclaim in impatience, " can you imagine for a single serious moment that the First Cause intended trout to be caught by the methods of sport, so that men might find pleasure in the capture of them ? "

I can imagine this, and purpose to explain why ; but in the meanwhile let us observe a most peculiar thing. In other times the dominant philosophy was what

is called anthropomorphic. It attributed
to the First Cause the character of man.
Now the dominant philosophy is at the
other extreme.   It affirms that the
utmost efforts of man, in his limited
processes of cognition, leave the First
Cause unknowable.   Thus it would
appear that our philosophy has under-
gone a complete revolution.   Has it
really ?   I think that the appearance is
deceitful.   As far as philosophy is con-
cerned, there is very little difference
between the assumption that the First
Cause is the prototype of man and the
apprehension that the character of the
First Cause is unknowable.   In both
cases we are presented with affirmations
about the First Cause ; both affirmations come from developments of the
human understanding ; both are crops of
belief on the gray matter of similar brains,
cultivated differently.   Each, that is to
say, is the human understanding egoistic,
asserting itself, controversial ; and arguing
with the same means from precisely the

same premisses, which are not knowledge
gained from the arguers having actually
witnessed the First Cause at the first
action in the void, but inductions about
the nature and purpose of that action
derived from study of its results millions
of centuries afterwards. Thus, though
they differ in their conclusions, Anthropomorphist and Agnostic Evolutionist are
identical in their methods. The finite
intellect of man is the basis of the belief
of each. One believes that God made
man in His own image; the other believes
that man is gradually developing into the
image of God. The Agnostic Evolutionist is the Anthropomorphist inverted.

Between the two there is a still more
striking resemblance. When thoroughly
in earnest about their philosophies, both
are inclined to be puritanical. Neither
has any rational cause to be so; but both
of them are. Neither can ever completely
divest himself of a feeling that the pursuit
of pleasure should be suspect. Pleasures,
both would say, are incidental to life,

which is a serious business : they should
not be regarded as an end in themselves : they are but eddies in the main
stream, upon the tendencies of which the
philosophers ponder gravely, awestruck.
Neither of the philosophers, as a rule, has
the artistic temperament, to which all the
world is interesting and good simply
because There it is, and wonderful ! In
both of them the sheer joy of living is
curbed and subdued by thought. In their
very effort after precision of detail and a
synthesis, they lose sight of the subject
they set out to study. Particulars, many
of them most suggestive, are lost in
generalisation.

Let us reflect on some of these uncon-
sidered trifles.

It is a sunny afternoon, and it would
be pleasant to be outside ; but I hear that
Tim the terrier is about. If I went to a
seat in the open air, I should have no
peace in which to ruminate and write.
Tim would come with his ball, a lawn-
tennis one, and lay it at my feet, inviting

me to throw it. When the ball was
thrown, off he would bound after it ; and
in a few seconds, short tail wagging and
eyes gleaming, he would be back with it
in his mouth, to be placed at my feet
again, in continuance of the game. This
would be repeated all afternoon, and the
day's task would be left unaccomplished.
The need to go on with one's work keeps
me lying low indoors, aloof from Tim's
importunities, to which, so great is his
disappointment when denied, I should undoubtedly have to yield. On the other
hand, Tim sometimes comes to the golf-
course. He walks round with the
players, and is obviously interested in
what they do ; but what does he see ?
He witnesses everything save the count-
ing. The one thing of which Tim is
unconscious is the one thing that would
make completely intelligible to him the
pastime of his human friends. May
it not be that the game of ball which
he makes me play with him would
be intelligible to me if only I knew why

he plays it? Perhaps he is challenging
me to throw the ball to a place where
he cannot find it; perhaps he has some
method of reckoning time, and wishes to
show in what a brief space he can recover
the ball and restore it. One can hardly
doubt that in his part of the performance there is something I do not
see, some consideration my ignorance of
which makes our game of ball as purposeless to me as the game of golf is by a
similar lacuna made to him.

If that is not certain, it is possible;
and the possibility will suffice for the
purpose at present in hand. It suggests
that many things in this world which are
so commonplace that they are usually
unheeded by men and women, and particularly ignored by those who seek comprehensive generalisations, may, far from
being purposeless accidents, be actually
vital parts in the scheme of creation.
Among these commonplace things, pleasures are conspicuous. For what do we
strive and slave? There is, of course,

compulsion in the certainty that if we
lack "independent means" we should
starve if we did not work; also, there is
the sense of duty to relations, friends, and
the Throne; but no one, I imagine, can
deny that a hope of leisure and the means
of enjoying recreations is invariably at the
back of toil.   It will be useless for any
one here to cry out "Hedonism!" as if
that would blow my argument into a
bubble to be pricked.   'Isms, as we
dramatically behold when we think of the
economic variety now that the Anglo-
Saxon race has reached positions which
no race ever held before, do not settle any
problem.   They are but the terminology
of a refined kind of wrangling, and more
often perpetuate errors than they destroy
them.   If pleasures be not the end of life,
it is difficult to perceive that life can have
a purpose at all.   Can this be gainsaid by
our own orthdox, whose constant hope is
heaven; or by the Mahommedan, who
gladly dies in battle because the way of
war is the sure path to paradise; or by

the Buddhist, whose view of the here-
after is but a variation of the hopes of
the European peoples ?   Surely it cannot;
and surely, also, if pleasure is in the
expectation of all theologians the supreme
quality of the ultimate life, it cannot but
have a natural sanction in the present.

This reasoning might be developed to
vindicate sport against the aspersions of
the many persons who feel that there is
something dubious, probably sinful, in all
not-absolutely-necessary actions that give
pleasure to men and women ; but that is
not at present needful.   I am not assuming that sports require defence.   I am
only assuming that one of them, angling,
may be made the more delightful by
being interpreted as something other than
one of man's many inventions.   It is an
invention, unquestionably ; but if the
argument about design in nature is not to
be wholly abandoned, the sport, it might
be held, has no less a sanction than that
of having been part of the creative plan.

Now, the argument from design always

involves, as the first step, an argument into it.

It matters not whether we view the subject in the light of the old orthodoxy, that of Genesis regarded literally, or in the light of the new, that of Evolution : thinkers of both schools agree that species were created not to be destroyed, but to be perpetuated. Well, if there is any species in marvellous harmony both with its own environment and with the desires of man, it is the favourite fish of our streams and lakes.

By a critical process of exhaustion, we have learned a good deal touching the life of the trout. We have seen that it is neither the wind nor the want of wind, neither the glare nor the gloom, neither the heat nor the cold, that puts him in the mood wherein, as a rule, he is safe against the assaults of his chief enemy, who is man ; yet we realise that, when all is said that can be said, there remains some undiscovered provision of nature protecting him against his own voracity.

Even as Tim in the game of ball is doing something which, if it were intelligible to us, would render his action manifestly intelligent, the trout is in some supremely providential relation to those vapours in the atmosphere which mysteriously keep him down.   Otherwise the fish would soon become a mere tradition in this populous land of sportsmen.

That is not all the marvel.   If the secret of the obscure atmospherical conditions could be detected and a means of undoing its influence could be discovered, mankind, as has been noted, would cease to be interested in the fish as a subject of sport.   It is just because the trout are difficult to catch that there is pleasure in catching them.   Thus, the very atmospherical influences which defeat our endeavours on the water are an indispensable condition of our enjoyment in the pursuit. It is only rare possessions and difficult triumphs that men prize.

Besides, is it not wonderful to realise that the methods of the sportsmen are the only conceivable methods by which trout could be taken without unnecessary suffering ? Imprisonment in a net would prolong their terror more than the sportsman does. Damming a stream in order to leave them defenceless on the dry bed below, or liberating the waters of a lake for a similar purpose, would entail the death of many more fish than were wanted, and could be frequently repeated only at the sacrifice of the whole stock. Painless death would follow a handful of dynamite hurled into a pool; but, whilst only a few of the fish in the pool might be wanted, all of them would perish. By throwing lime into the water, you could easily poison a stream for miles. On the other hand, the methods of capture adopted by the sportsman seem to be exactly in accord with the balance of nature. They prevent overstocking and degeneration; and, as general experience shows, they do not unduly reduce the numbers of the trout. Then, what of the sportsman's methods ? Was it only by a

blind accident that the First Cause, providing that trout during the months when
they are fit to be the food of man should
feed liberally on flies, provided also that
man, who, in common with all animals, is
instinctively a hunter, should find feathers
and other materials out of which exact
imitations of the flies could be made?

This could be affirmed only on the
assumption that the First Cause lacked
foresight.   If the imitation flies were
foreseen, they were ordained: in the
Omnipotent foresight and ordination are
indistinguishable.

This thought is open to the objection
that it would make the First Cause
responsible for the evil that men do as
well as for their harmless or praiseworthy
actions.   The Omniscient, that is to say,
must have foreseen evil as well as good,
and, if foresight and ordination are
ultimately indistinguishable, must have
ordained evil as well as good.   Thus,
clearly, the argument from Design is not
a complete explanation of the universe.

It stops short just where successful progress into the past would begin to render it absolutely convincing. Still, as far as it goes, the Argument is held in general respect, and it may not be profitless to pursue for a few sentences farther its bearing upon our subject.

It might be said that in relation to the trout the Argument from Design, even with the modification which we have seen to be necessary, would break down completely if the people suddenly resolved upon abolition of the Game Laws. This would exemplify a strangely persistent error of the human understanding. There is nothing that we know of to render impossible a snowstorm that would blot out the whole of the peoples of Europe; yet the snowstorm does not come. Twenty thousand citizens of London marching against St. Paul's would, by the impact of their own mass, bring the Cathedral to the ground; yet the march is not undertaken. As some men die by their own hands, it is conceivable that

11

nations might so die ; but nations never do.   Similarly, there is never a protest against the Game Laws sufficient to bring about their repeal.   Clearly, then, the preservation of game is as directly referable to the scheme of creation as are the preservations of humanity in Europe from the snows, St. Paul's Cathedral from the mob, and nations from the impulse to suicide.   It is only to those who think in 'isms, or do not think at all, that this statement will be startling or incredible. All things in this world are wonderful ; and sometimes familiar things, seen in their true relations, are the most wonderful of all.

Often, however, it is very difficult to perceive the true relations.   This is notably the case in human society.   Some social phenomena are more puzzling than any to be witnessed among the lower animals.   As far as one can make out, these do not habitually do anything without a cause ; but men are different. Dogs, for example, never bark merely for the sake of barking; but men frequently speak merely for the sake of speaking. Even as the literary style of gentlemen who despise syntax is full of unrelated participles, the colloquial style of others is full of things that have no perceptible connection with reason.

Of this I had a striking series of illustrations on returning to Town after a long absence in an almost uninhabited land. In St. James's Street, on my way to the Club, I met a man, and he said, "There's air!" "Doubtless," I answered, without understanding. My friend was a very well-known artist, none other than Mr. C—— W——, who, after the exchange of a few words on the weather, words of a more definite kind, passed on. Immediately on his going, I met another man, a man for whom I have a very high regard, Mr. J—— A—— G——, head of a department in Science and Art at South Kensington, who greeted me with an astonishing statement. "What ho! she bumps!" quoth he. Not comprehending, I changed

the subject, and asked whether he had
taken any steps concerning the seat
in Parliament towards which I had long
been urging him. He invited me to
lunch at his house three days thence, as
of yore, and went his way. Before I
reached my destination, not more than
two hundred yards off, the strange an-
nouncements were repeatedly insisted on.
Lord A—— O—— and his rival in
foreign travel and adventure Mr. T——
C—— positively assured me that there
was air ; and just as I was turning to the
steps of the Club, Mr. M—— W——
thrust his beaming countenance out of a
hansom cabriolet, and shouted the tidings
that she bumped. "O, Monty, kindly
go to the devil ! " I implored, entering the
stately asylum. It turned out to be an
asylum in more than one sense. Old
habit led my footsteps towards the round
table in a corner of the coffee-room, and,
seeing welcoming faces there, I sat down
for luncheon at the accustomed place.
The talk was about politics, and much of

it I could follow ; but there was one constantly recurring word of which I could make nothing. "Efficiency," "Efficiency," " Efficiency." It was sprinkled over the dialogue of my much absorbed companions, and from all the tables in the room the earnest sibilants penetrated the cheerful chatter of the mid-day meal. In the smoking-room shortly afterwards I narrated to Mr. G—— B—— the strange things I had heard, and asked whether he could explain. His answer did but darken counsel. It was in music. Lifting up his voice, G—— B—— chaunted : " Some one | ought to | speak to | Mister | Hodgson | Some one | ought to | tell him | what to | do-oo ! " " Evidently there are rogues about," I said to myself, moving off towards a shady corner in which I had espied Dagonet, in an armchair, meditatively flourishing a large cigar. Dagonet is an encyclopædic Briton, and very humane : I dared say he could and would explain the words that had puzzled me on my return to Town.

" What's air, Dagonet ? " I asked.

" Air—the word, that is—comes from the root אור, *aoor*, Hebrew and Chaldee ; which means, *to shine*. The sense is *to open, to expand ;* whence *clear;* or *to flow, to shoot, to radiate*. Air—the thing, I mean — is inodorous, invisible, insipid, colourless, elastic, possessed of gravity, easily moved and rarefied and condensed. In short, my boy, air is the fluid which we breathe to live."

"Quite so ; but what does a fellow mean when he says, ' There's air ! ' ? "

"Obscure in origin," said Dagonet gravely.   Certain philologists hold that the words were uttered by Mr. Gladstone when he first gazed upon the atmospherical amenities of Blackpool, or those of any other holiday resort anxious to have itself made dear to the people."

" Well ; but why do men keep firing the words at you in London ? "

Dagonet laughed, and unbent.

" O ! they're a mere catch-phrase of the Town.

" And She who Bumps—what ho ? "

" She's a phrase of the Town too."

" And Efficiency ? "

" That's in the same category, sonny."

" And all these phrases—do they mean anything ? "

" Nothing whatever. But the carriage will be at the door. Come on to Lord's —Oxford and Cambridge."

If you are thinking on some subject which does not involve serious practical interests, a drive through London at the height of the season is a great help. The bright, gay bustle of the Town is not a distraction : it is at once soothing to the nerves and an impulse to meditation. Thus it was that certain things appeared to me in a new light as Dagonet and I were driven to the cricket match.

Dogs, as must have been perceived from what has been told of Tim the terrier, frequently attract your curiosity by actions which, though inexplicable, would probably be rationally accounted for if the animals could speak. Men, it

had become clear, attract curiosity by a
proceeding exactly opposite. Frequently,
in set phrases as definite to the ear as the
dog's gambols are to the eye, they speak
when they have absolutely no meaning to
convey. At other times their words are
dogmatic and challenging in proportion
to the haziness of their ideas when they
really have some. Of the one set of
phenomena examples had been provided
in the phrases about the air and the lady
who bumped. Of the other "Efficiency"
seemed typical. It was a word too big for
the doctrine it enclosed : a balloon, as it
were, without gas enough to float it,
striving to rise, but constantly falling,
unshapen, to the earth : a symbol without
a substance : a political ideal exciting to
many persons until they should have had
time to discover its close kinship with the
primitive, simple-seeming maxims, such as
" Men are born free and equal," which in
all ages inspire eager minds before the discovery that nature is too picturesque to be
perfect, knowing nought of the ideologue's

crude precision, and almost insolently un-
democratic : Even when we consider the
mountains insignificant, the earth is but
approximately round, and it was not born
either equal to any other globe of the
solar system or free to run in the heavens
a course of its own choosing.

Could it be that " Dry Fly " merited a
place in the class of phrases which had
just been adorned by the tumultuous
advent of " Efficiency " ?

This dark question had presented itself
about a year before. At that time, as
occasion arose, I was contributing an
article now and then to *The D——
C——*. Mr. H—— W—— M——, the
Editor, who is respected and beloved by
all who know him, especially by such as
have humane Tory insight, had done me
the honour to invite writings. Being in
ignorance about the solidarity of man and
the trend of progress, I was never to
pen a word touching politics ; but on
such subjects as are afforded by field-
sports, subjects of purely pagan interest,

I was to say what there was to be said
quite frankly. This would help to show
the world that, although *The D——
C——* was the friend of Humanity, it was
not the enemy of man. It had still
some lurking self-consciousness of the
original sin that makes all of us ready
for the chase, and would not be ashamed
of bearing symptoms of this undeniable
predisposition to Anti-Jacobinism.

All went well for a good long time,
and I quite enjoyed my charge of the
Nonconformist Conscience in its sinful
tendencies. I felt like a high priest of
Satanism in a new and rather rational
school. Trouble, however, was at hand.
One evening H. W. M., pale and obviously perturbed, came to speak to me as
he was passing out of the coffee-room.
"That leader this morning," he said,
haltingly,—"H——, who has just been
dining with me, says it's something awful."
"Ah! I'm sorry, Editor; but why?"
"Well," said H. W. M., taking a seat
very seriously, "H—— says you're all

wrong about the Dry Fly ; that you do
not seem to understand what Lord G——
was trying to say in its praise ; and that
it is a pity we treated his book just as if
we thought that because a man's a Tory
he can't be right even about trout-fishing.
I must say I agree with H—— "

In a way this was pleasant hearing.
Mr. L—— V—— H——, who had been
arraigning the article in H. W. M.'s
journal, is the rising son of a great Liberal
family, and it was chivalrous of him to
protest when he considered that a political
opponent had been ill-used. It was
necessary to admit, also, that I had not
viewed Lord G——'s book quite without
what might be considered prejudice. For
a good many years the Dry Fly had been
a craze. Writers in the journals of sport
were always penning delirious rhapsodies
about it ; and the very ladies you took in
to dinner, most of whom did not know
one fly from another, enlarged upon the
subject. Lord G—— had gone even
further. Not content with expressing his

liking for the new method, he had sneered at anglers who used the old method, which he flouted as "the chuck-it-and-chance-it principle." In the literature of the open air, his style had jarred. It had seemed to me that the Lord Chancellor would not have been more incongruous if by way of introduction to the Speech from the Throne he had danced a reel on the Woolsack. I submitted those thoughts to the Editor. Surely, I urged, he would allow it to be a matter for grief that Lord G—— had been lacking in urbanity? Angry words were a misfit in the literature of sport. A game-shot was not derided because he didn't have a Purdy gun : why should an angler be jibed at because he didn't use the Dry Fly? The Editor smiled; but he was not persuaded. His answer was to the effect that the article had been written in a bit of a temper too : that the writer had been vexed with Lord G—— because his style had not been to the credit of Toryism, which, it seemed, was a matter of general taste as well as a matter of specific opinions; and that a main purpose in the critical attitude towards the book was to arrest the downfall of Toryism by preventing the spread of bad style. A fine purpose to put *The D—— C——* to! Reactionary propagandism in disguise: jesuitical, it might almost be said. But to the cold facts, H. W. M. continued, speaking with gentle impetuosity: Was the Dry Fly right, or was it not right? "Does H—— say that it is the best way of catching trout?" I asked. "He says more than that," the Editor answered: "he says that on chalk-streams it is the only way."

The discussion might have gone on rather aimlessly but for a fortunate message from T—— J—— B——, who had friends at dinner across the room. The port was very good: if H. W. M. and I had finished our meal, wouldn't we join him? We did; and B. asked what was the trouble we had been considering so earnestly ; and the Editor spread the dilemma out. " O, that's easily settled ! " said B., who is ready in resource. " We'll put the question to the proper test. We'll have a trial of the Dry Fly against the Wet Fly. Mr. H—— will use the one, and the writer of the article will use the other. Stakes, ten guineas,—to go to the Open-Air Fund for Children. I back *The D—— C——* view for another ten — bets also to go to the fund." The Junior Member for W—— indicated readiness to accept the wager. " Agreeable ? " said H. W. M., still anxious, turning to me. " O, Editor ! what do you take me for ? " I answered. At these words the Editor's expressive face lit up with reassurance. " Well, where's H—— ? " asked B. " Gone to the House of Commons," answered H. W. M. " Could you see him to-night ? " " Perhaps ; but he is very anxious to hear the debate on the Budget." " O, never mind that ! " said B. " If you're looking in at the House, haul

him out and arrange. Will you ? " The
Editor vowed he would rejoice to do so.
"Good," said B. "I put my stream at
your disposal for the match, which might
be this coming Saturday if you're all free,
and H—— is free ; and you'll all be my
guests at the old inn. Let me know by
telegraph to-morrow if H—— can go.
Then I'll get Senior of *The Field* to be
umpire."

Soon afterwards, his misgivings about
the angling policy of his journal much
modified, the Editor took his leave in
high spirits ; and next morning there was
notification that Mr. H—— had been
found willing. A leaded leading-note in
The D——— C——— stated that, serious
objection having been taken to the article
questioning Lord G———'s doctrines about
Fly Fishing, it had been arranged that
the objector and the writer of the article
should put the question to a test by
angling on the same stream, a chalk-
stream, on the same day. The one
would use the dry fly ; the other would

use the wet. The result would be announced on Monday, and, whatever it was, the Open-Air Fund for Children would benefit by the entertaining and instructive incident.

Alack, the project was not quite accomplished. The night before the eagerly expected Saturday, Mr. H—— sent to our host a note saying that in accepting the invitation he had forgotten an engagement to entertain guests, "a Dry Fly party," on his own stream that very day. This was unfortunate. A carrying-out of the arrangement might really have shown the unreasonableness of debating in anger the principles of a peaceful sport. However, the plan did not exactly come to nothing altogether. The Dry Fly and the Wet Fly were tried on the same day, and it chanced that I met H—— shortly after his return to Town on Monday. His basket on the Saturday had been one of fifteen trout which weighed twelve pounds. The basket on B.'s stream which represented scepticism about the Dry

Fly doctrine contained twenty-five trout weighing thirty-three pounds. One of them, taken on a Greenwell's Glory, was five pounds and a half. The two baskets could not be regarded as affording grounds for a satisfactory judgment on the question that had been under discussion. H——'s stream may not have been so good as the other, and the atmospherical conditions may not have been identical. Still, a basket of thirty-three pounds was sufficient to persuade H. W. M. that *The D—— C——* had been justified in its protest against Lord G——'s unnecessarily earnest derision of the ancient method.

On a subsequent occasion, when again the wet-fly basket was not despicable, Mr. Senior remarked, "Yes: I admit it is good, even surprising; but I am quite confident there are waters where this could not be done." Where are they, I wonder? Once another scientific fisherman, Mr. A—— L——, took me to the Test in order to see whether there

12

was any truth in the reactionary heresy
against the much-extolled Dry Fly. He
caught eight trout, each a little above
three-quarters of a pound, the limit on
the stretch which we were fishing; his
friend Mr. C—— also had eight, of
similar size. The wet-fly basket was
twenty trout of the same dimensions.
Once, at the invitation of Sir W——
P—— and Mr. W—— M—— R——,
I had a very pleasant day on the
Kennet, the trout in which are generally
supposed to be proof against all flies but
the Mayfly. At the end of it the creel
held sixteen trout weighing over twenty
pounds. The only very well-known
English river on which I have not fished
is the Itchen, and I cannot easily imagine
that water to be wholly different from
others on which the old-fashioned method
seems still to be not without merit.

Being averse from such a narrative of
my own experiences, I would strike out
the last two pages if that could be done
without impairing the argument; but if

one is alone in a heresy, which at present
is apparently the case, how is the truth to
be arrived at unless the facts on both
sides are revealed ? I have no vanity in
the brief record which has just been
penned. Indeed, it seems scarcely less
out of place than the language on the
part of Lord G—— which led to the
discussion ; and it can be justified, if at all,
only on the consideration that when one
has a theory on our fascinating subject it
is well, if possible, frankly to support it
by statistics. All I mean to suggest is
the possibility that even in sport, an
activity of the daylight and the open air
the human mind is liable to become the
victim of a phrase.

In an early chapter I have briefly set
forth one of the reasons for believing that
trout often feed upon drenched flies. At
other times, it is certain, fluttering flies
have their seemingly exclusive attention.
Then it is that the dry-fly man finds
his opportunity, and I should be the last
to deny that it is very inspiring. To

most of what has been written about the
delights of " stalking " a rising trout
one can give unreserved assent. Every
moment of the action is peculiarly aglow
with the spirit of the chase. In a
manner which is telling from its very
simplicity, this charm, so enthralling in
itself and so difficult to reproduce in
words, has been expressed by Mr. R. B.
Marston. Recounting an afternoon on
the Tweed, he wrote to Mr. E. M. Tod,
who published his letter : " In about
" three hours I killed a nice basket of
" over twelve pounds of trout, all with
" the fly, and quite two-thirds with the
" dry fly. I used your double - hook
" midges, three on my cast (Greenwell's
" Glory and Iron Blue did best). I fished
" all three flies first dry and then wet. I
" also fished with two of the flies dry and
" one wet, or one dry and two wet, and
" this in the rapid broken water of the
" streams as well as on the pools. It is a
" great mistake to think dry-fly fishing
" must be confined to slow smooth water.

"Wherever the natural fly can float,
"there the artificial can float if properly
"made, and oiled, and used. It is most
"interesting to watch your fly coming
"down dancing on the waves, and then
"disappear when the brown head of a
"trout breaks the surface; also to see it
"pulled under when a trout takes one of
"the wet flies." Once I was witness of
the same feeling expressed without aid
from the literary art. I was casting into
a pool in a Hampshire stream. "See!
see!" some one behind me exclaimed.
I turned; and there, in his shirt-sleeves,
was the landlord of the little inn at
which I was staying. With outstretched
arm he was pointing to something in
the blue air athwart the copse bordering the water, and his eyes were gleaming with some sudden joy. I looked
towards where he pointed. It was
the first Mayfly of the year that had
moved him. The innkeeper was a
reserved, shy man, who at ordinary times
could scarcely be induced to talk at all;

but that fluttering Mayfly, symbol of
summer at the noon and all the green
world at the freshest intensity of throb-
bing life, had stirred him to a panic
happiness. Now, something of the same
joyous emotion comes at the sight of
a fly, with cocked wings floating, that
has been lightly cast where a great
fish is known to be on the look-
out. This I know full well. Still, as
we are endeavouring to treat the whole
subject in a scientific spirit, it is necessary to point out that the delight is
not always unmitigated. Sometimes a
trout takes the floating fly; but how
often does he rise and miss? In my own
experience missing is the rule. Up comes
the trout, and down he goes; but the fly
is where it was, on the surface. It is not
that I have missed the fish. It is that
the fish has missed the fly. This is very
often what happens on a river, and it
almost invariably happens on a lake.
Were it not for an astonishing fact, which
I will mention immediately, it would

argue wariness on the part of the trout.
One might believe that when a trout has
risen at a floating fly and gone down
without it he has detected or suspected
the thing to be a lure. It is pleasant to
think so, for much of the fascination of
the sport is derived from the feeling that
human skill is matched against astuteness
in the fish ; but now I fear that another
tradition must be sacrificed, or at least
modified. If the trout suspects the
artificial fly, he is equally suspicious of
the natural. Day by day, as I write, I
have been watching his behaviour carefully. It is not only my fly that he
usually misses. He misses the real insect
as well.

Has it been generally noticed that
there are at least three different kinds of
rises ? There is the rise, in leisurely
manner, which is, as it were, finished off
by a slow wave of the trout's tail above
the water. That, though I cannot make
out why it should be so, is when, early in
the day, there are on the water myriads of

minute black-and-white gnats upon which
the fish are feeding. Then, there is the
ordinary rise, when, if it be at a natural fly,
the trout just tips the surface and retires
without anything like a somersault, and
when, if it be at an artificial fly, it is a
business-like swift action without fuss.
That, I think, is when the fish are feeding
on insects slightly below the surface.
Again, there is a rise which is hard to
describe but beautiful to see. No part of
the trout is visible; but he must have
been very active for an instant. Swiftly
the water breaks, swirling as the ripples
rapidly expand, in a manner quite different
from that of the ordinary rise, which is
usually but a slowly-spreading dimple.
That is a few minutes after a hatch of the
larger insects. The trout do not move
when the single spies appear; but when
the battalions are abroad the movements
are rapid and exhilarating.

Surely this, if ever there be one, is the
time for the floating fly ! Of course it is,
and I do not neglect it; but it is

necessary to confess that my best efforts
are almost always of little avail. Not
only do the trout miss my own flies : they
constantly miss the real flies. Sometimes I see one taken ; but much more
frequently the insect is still afloat on the
swirl as the trout goes down. By and
by, when the rise of flies has gone on for
a time, or when the whole hatch has been
on the water for half an hour or so, I find
fish with a cast of flies slightly sunk.
Why ? The obvious explanation seems
to be that, although the trout begin to rise
soon after the first risings of the fly, they
do not begin to feed in earnest until
many of the flies have been drenched.

Reasoning thus after many days of
observation on lake and stream, I thought
it would be well to examine methodically
the literature of the Dry Fly. Surely, I
felt, there must be some scientific consideration, which I had completely over-
looked, to account for the practically
unanimous enthusiasm with which the
anglers of England had accepted the

theory that artificial flies should float?
Well, I obeyed the judicial impulse; and,
after diligent search, I came upon relevant
evidence which was surprising. The
passages presenting it were these:—

"No doubt the Salmonidæ in rivers
"will at times take, and take freely,
"winged flies on the surface; but, besides
"minnows and other small fish, Crus-
"taceans and Molluscs, their staple food
"consists of Caddis or larvæ of Trichop-
"tera, and the larvæ of Ephemeridæ,
"Perlidæ, Sialidæ, Diptera, and many
"other land and water-bred insects.

"As one of the few fishermen who
"have for many years consistently
"studied the food of the trout and
"grayling by the only available and prac-
"tical means, i.e. autopsy, may I be
"allowed to tender my evidence? I have
"invariably found that the undigested
"insect food has consisted of masses of
"larvæ and nymphs, with a few occasional
"specimens of the winged insects. This
"has been the universal result, whether

"the trout or grayling have been taken
"in waters fished daily, or in compara-
"tively wild parts where they seldom see
"an artificial fly. In rivers where in the
"memory of man no stocking had taken
"place, or in others, which, from neglect
"or other causes, had been depopulated,
"and where, therefore, a fresh generation
"of trout had been turned in from the
"pisciculturist's ponds, the experience
"has ever been the same. The earliest
"autopsies taken do not differ at all in
"this respect from those of the latest
"date.

    "In the case of certain of the Ephe-
"meridæ 'either the mother alights upon
"'the water at intervals to wash off the
"'eggs that have issued from the mouth
"'of the oviducts during her flight; or
"'else she creeps down into the water
"'(enclosed within a film of air, with her
"'wings collapsed so as to overlie the
"'abdomen, and with her setæ closed to-
"'gether) to lay her eggs upon the under-
"'side of stones, disposing them in

"'rounded patches, in a single layer
"'evenly spread, and in mutual con-
"'tiguity.' After laying her eggs she
"floats to the surface and flies away, un-
"less perchance her setæ or wings have
"become sodden, in which case the brief
"remnant of her life is sacrificed to her
"care for the next generation. Every
"observant fisherman has at times, when
"wading, been surprised to find a number
"of spinners crawling up his stockings
"and brogues. Doubtless these are the
"females striving to regain the surface
"after depositing their eggs in the
"manner just described."

In a very scientific way, two things
were thus put beyond question. In the
first place, instead of being the main food
of trout, flies are only an occasional
luxury. In the second place, besides
being liable to fall on the water and be
drowned at the coming of strong winds
or of untoward chills, all female flies, in
the course of nature, go down into the
water voluntarily. Is it to be supposed

that the trout disdain them as they go, or as they come upwards when the eggs are laid ? Is it not much more reasonable to suppose that it is then, when the flies are under the surface, that the fish indulge in their occasional luxury ?

This thought is strikingly supported by the fact, on which I have dwelt, that, although flies on the surface attract the trout to rise, they are often left there when the fish have gone. Indeed, the evidence I have quoted is practically a complete scientific demonstration that, if the purpose of angling is to catch trout, the Dry Fly doctrine, far from being in accord with the teaching of Nature, is flatly repudiated by the all-wise Dame. The lures should be allowed to dip below the surface.

The evidence is from no dubious source. The quotation within the third paragraph is from the writings of the Rev. A. E. Eaton, of the Entomological Society of London, who is described as "the first living authority on the Ephemeridæ";

and the whole of the evidence transcribed is from Mr. Frederic M. Halford's *Dry Fly Entomology.*

Thinking on what I had read in the stately and authoritative volume, I recalled a picture in a book on Angling. The author has crept on hands and knees towards a pool in which there is a rising trout, and is in the act of throwing over it a dry fly. Does this sportsman, so earnestly expounding the fashionable doctrine we have examined, know the illuminating confession of a certain barber familiar to all who frequent the neighbourhood of St. James's? "O, yes, sir," said the barber confidingly: "the lotion certainly does good; but it does so in what may be called an indirect manner. You will see that the instructions on the bottle say, 'To be well rubbed in.' The truth is, sir, it is not the lotion, fragrant and cooling as it is, that does the work: it is the rubbing in." Similarly, if our very serious fisherman in the picture catches the big trout, he will owe his

success not to the consideration that the
fly has floated, but to the consideration
that he is crouching and out of the fish's
sight. Thus far, and no farther, the Dry
Fly is an unimpeachable counsel of
perfection. Less than thus far, like
"Efficiency," on which I have touched
because it illustrates the same entertaining phenomenon of natural history in
another domain of thought, it is merely
one of those phrases which are so
strangely attractive to the masses of men,
and so contagious : a symbol which, while
enchantingly revealing the ideal half of a
truth, conceals the other half, in which
the realities, the hard facts, lie. There
are not only microbes that afflict the
body : there are also microbes that afflict
the mind : and, just as pollen, the fecun-
dating dust of weeds, is distributed over
the hospitable earth from the wings of the
flighty bees, the microbes of the mind,
the half-truths, inspiring symbols, are
planted in the hospitable emotions of men
from the wings of words, words, words.

CHAPTER VII

LAKE AND STREAM

WHICH is preferable, lake or stream?
That is a question which one sometimes
hears discussed; but I have never heard
it considered in relation to the real

contrasts which the two kinds of water present. Some men prefer the stream, because there they are obliged to walk, and walking, on a holiday, is pleasant; others prefer the lake, because there they sit in a boat, and that is soothing after the bustle of business. These, however, are casual thoughts. The fascination of angling lies largely in the problems of natural philosophy with which the sport is fraught, and these can be but imperfectly understood through an acquaintance with one kind of water only. The trout in a river, it is true, are pretty much the same as those in a lake; but that in itself is a surprise. It might be expected that the one tribe of fish, which have to be constantly in exercise against a current, would be stronger than the other, which are habitually at rest; yet that is not the case. The lake trout are just as game as the river trout. I think, too, that the various atmospherical conditions have the same influences on the trout of the stream as they have on those of the

lake. Only, at least two of these conditions are modified on the stream. I
allude to the temperature and the wind.
Flowing water, naturally, is of the same
temperature all through ; and, tumbling
over a fall here and there, it is frequently
aerated, which in some measure neutralises
the excess of noxious vapours with which
the atmosphere is now and then charged.
For those reasons, sport is less markedly
inconstant on rivers than it is on lakes.
If on a murky day dire results were certain to flow from failure in the attempt to
bring home a brace or two, I should
certainly prefer a stream.

On the other hand, the lake has an
attractiveness of its own. At many
places on most rivers you can actually see
the trout you may possibly raise. This is
true even of the Thames, concerning
which, now and then, we read in *The
Times* that a very fine trout has taken up
his quarters at Sunbury, or at Datchet,
or elsewhere, just as if he were some
great lady entered into residence at her

Town house for the season.   The trout in
lakes live in much greater privacy.   It is
only on rare occasions, as when looking
from a boat into a sand-bottomed bay on
which a rock reflects the sunlight, that
you can see any of them at all ; and these
do not tell you much.   They may be big ;
but they are no evidence as to the size of
the trout in other places.   There may be
a good many of them ; but that gives no
cause for believing that fish are in equal
numbers all over the water.   Thus, in
fishing on a lake, you never know what
your luck is to be.   Any day may bring
you a trout so big that the basket would
not hold it.

To most of those who habitually fish
with flies such good fortune comes but
rarely.   On certain Irish loughs very
large trout do sometimes rise at flies ; but
that is in exceptional cases.   They rise
when the Green Drake, an insect of the
Mayfly family, is abroad ; but on most
lakes flies of that kind are never to be
seen.   On most lakes, therefore, the great

trout lie low.   Early in the year, now
and then, one of them, but hardly ever of
the biggest, does take a fly; but as the
season advances it is noticeable that the
size of the captured trout gradually de-
clines.   I have often wondered what can
be the meaning of this.   It would seem
either that a lake trout needs less susten-
ance the larger he grows, or that the larger
he grows the less does he care for flies.

Perhaps a gradual loss of appetite for
flies is the more natural explanation.
This is suggested by the fact that, whilst
they ignore the daintier lures, the large
trout will almost any day of the season
fall ready victims to a well-spun minnow.
Sport by that means is not to be despised.
Many of the trout which a minnow takes,
though large, are not old.   Most of them
are small of head and big of tail, shapely,
firm, and astonishingly brilliant in colour.
They fight with great vigour, and are
manifestly in the prime of life.

Some say that fishing with a minnow
calls for no thought; but that is a mis-

taken view. Who has not noticed how cunningly the experienced boatman, when you are trolling, goes about the business? Do you mark his course? It is not in a straight line that he moves: that would disturb all the trout over which, following the boat, the minnow would pass. Therefore, instead of going straight, the boatman pursues a line which is in large curves, curves such as some giant must make when he cuts the outside edge and the inside edge alternately by the same leg while skating on the ice of Lilliput. By this means, the gillie contrives that the minnow, which is about a hundred yards off, shall cross the path of the boat only now and then, and, for the rest, be moving through water that has not been disturbed. Though simple, it is a cunning plan, showing that fishing with a minnow calls for thought; but is the thought in this case sound?

Doubt arose from noticing that frequently, when one was rowing or being rowed by short cut to the beginning of

some new drift on a lake, a trout rushed
at a fly trailed behind the boat.   If the
passage of a boat scares the fish, how
does that happen ?   The answer, I think,
is twofold.   In the first place, there is
some cause for believing that the trout in
lakes where boats are frequent become
used to seeing the craft and are not much
disturbed by their passage.   Once on
Lochleven a trout just in front rose at
a fly and missed.   Almost immediately
thereafter, the boat drifting rapidly, I
cast, in the teeth of the wind, behind,
raised the trout, and caught him.   Of
course, it is only an assumption that it
was the same trout ; but the reasonable-
ness of the assumption is very great.
Incidents of that kind are plentiful
enough to afford ground for believing
that the fish are not scared by the passage
of a boat.   In the second place, I am not
sure that it is only the trout by the very
eyes of which the minnow passes that
are attracted by the lure.   It is necessary
to remember that, as mentioned near the

beginning of this book, the trout, like the
salmon or the pike, seizes your minnow
because it seems to be a minnow wounded
or in trouble. Like the salmon and the
pike, the trout, taking not the slightest
notice of whole shoals of minnows sound
in wind and fin, will greedily, or cruelly,
or obeying some law of nature, probably
the one directed against the survival of
the unfit, rush at a minnow which appears
to be suffering in some way. Well, then,
is it not extremely probable that the fish
which takes your trolled lure has rushed
at it laterally, from a good distance
off? I imagine so, and the incidents I
have mentioned support the surmise ; and
if I am not wrong our serpentine boatman is strategic in sinuous error. One
would fare just as well if he pursued the
straight course.

Nevertheless, even if we have to
abandon the belief that it is really the
gillie, by virtue of his wariness, that is
the sportsman in trolling, there is still
much scope in minnow-fishing both for

knowledge and for skill. Minnows are of
considerable variety, and the trout are no
less particular about them than they are
about flies. Sometimes they will look
only at a blue minnow, sometimes at a
brown one, sometimes at a green one,
sometimes at a gray one with a scarlet
belly, sometimes at one which is all of
silver hue, and sometimes at one which
seems to be made of clay. At times they
will be rather indifferent to any or all of
them, and take an Alexandra fly. If
you look at an Alexandra in the water,
you will see that the feathers of which it
is made shrink and close, becoming com-
pact, instead of tending to expand, as do
the wings of most ordinary flies. The
Alexandra, therefore, is not really a fly: it
is a minnow in disguise. This seems to
have been discovered by certain makers of
tackle, who now openly busk the peacock's feathers on a triad of long-
shanked hooks, with a swivel at the top,
and call the result the Halcyon Spinner.

With all these minnows to choose

from, and sport depending on the proper
choice, who shall say that trolling in a
lake is not a matter involving knowledge ?
It would be very difficult to give a complete account of the different minnows
and the times to use them ; but there is
one good rule. The artificial minnow
most likely to be successful on any day
is that which most closely resembles the
minnows swimming about at the edges of
the lake, specimens of which can easily
be taken in a close-meshed net, or in a
trap, or even on a small hook baited with
a worm. Better still is it to fish with the
minnows thus caught themselves : it is a
peculiar fact that, whilst trout seem sometimes to prefer artificial flies to the insects
which the artificer has imitated, they
always prefer a real minnow to one made
of canvas and paint, or of steel and paint,
or of peacock's feathers. So do the
salmon and the pike : only, in their cases,
not minnows only, but also small fish of
many kinds, including parr and trout, are
among the lures in trolling.

Used in another way, cast deftly into some deep pool from which the angler is screened by bush or rock, the minnow is a deadly lure on streams; but it is generally objected to there, and, I think, rightly. The minnow used in lakes is capable of defence on the consideration that the great trout there do not rise at any known fly. Used in streams it cannot be justified by such a plea. In flowing water the largest fish are admittedly slow to rise; but they do rise sometimes, occasionally with astonishing freedom, and it is proper that they should be reserved for those who use flies only.

Does the same argument condemn the worm on streams? For that purpose it is used, I know, and on many waters in England the worm is forbidden; but that seems rather a pity. Worm-fishing on a clear stream is not coarse work at all. To any but the very expert in the management of rod and line, success in it is almost impossible. On a typical chalk-stream, to throw a fly properly is

difficult enough : it almost appears that
the trout have eyes in their tails : it is dis-
concerting to note how they sometimes
scuttle off just as you think you are
within casting distance.   They are at
much greater advantage when it is a
worm, instead of a fly, you have to
throw.   You need just as long a line,
usually, in the one case as in the other ;
and a long line weighted with a bait that
is easily jerked off is very difficult to
control.   Indeed, the skill called for by
worm-fishing is so great that the streams
of England would not, I think, suffer
much by withdrawal of the prohibition.

Lest this should happen, let us consider
for a moment the evolution of the gear
used in worm-fishing.   It is a remarkable
instance of how slowly, amid normal conditions, the inventive faculty of man
habitually works.   In days of yore, until
the time when, for example, Sir Walter
Scott roamed along the Border streams,
the worm was impaled on a hook which,
if the wire had been stretched straight,

would have been about two inches long.
Then arose an original thinker, Mr. W. C.
Stewart, to proclaim a better way. It
was surely obvious, he reasoned, that the
bent and rigid appearance of a worm on
such a hook must render the trout
suspicious. A free worm in the water
might not be always straight. It would
wriggle. Still, its general aspect would
be more nearly straight than curved.
Accordingly, Mr. Stewart invented the
tackle, a flight of three little hooks
whipped to the gut one above the other,
with a small space between the first and
the second, and another between the
second and the third, which made his
name famous among fishermen. The
upmost hook was slipped through the
worm at the head, the second at the
middle, and the other towards the tail.
It was held that on this tackle the worm
had a less unnatural appearance than it
could have when impaled on a single hook.
That was true; and thenceforth every
fisherman in the land, or at least in those

regions where the waters were not reserved
for fly-fishing exclusively, carried Stewart
Tackle in his book.

Without number were the published
praises of the gear, which was regarded
as perfect for nigh two generations.   Then
it began to dawn upon certain sportsmen
that progress in the art of worm-fishing
was still possible.   Why should there be
three hooks on the flight ?   Would not
two suffice ?   So one of the doubters
asked himself.   The result is that you
may now have " a new form of Stewart
Tackle " named after Mr. Cholmondeley
Pennell.   The only considerable difference
is that it has one hook less than the old
tackle.   In respect that the dark shanks
of only two hooks, instead of three, pro-
trude from the bait, the new gear certainly
is an improvement ; but why stop at that
refinement ?   Why not abolish another
hook, leaving only one, and that small ?

Before Mr. Pennell was working on
the problem, or perhaps during that time,
this question occurred to myself.   It did so

because I had noticed that a trout taken on Stewart Tackle was nearly always caught by only one of the three hooks, and that the upmost one. This brought to mind a statement that the trout invariably seizes its prey by the head. It suggested that the only use of the two other hooks was that they might possibly catch by the outside of the mouth a trout that missed or managed to eject the first hook. Otherwise considered, they were rather worse than unnecessary, to some extent hindering the lively movement of the lure, and at the same time showing, as it were, the cloven hoof. Thus, a small single hook in the head of a worm should be sufficient, and the bait should be the more attractive in that it would be almost untrammelled. On being put to the test, this reasoning was justified beyond expectation. The trout came very readily; and, still more gratifying, the single hook, so small that it could scarcely be seen when baited, almost always held.

On the English chalk-streams, the

limpidity of which is not much affected
even by a heavy flood, this simple tackle
affords delicate and exciting sport; but
there are many fine streams of quite a
different kind.   Different are all those
which flow through regions where brown
earth is ploughed.   These are the
streams in the Lowlands of Scotland and
of Ireland and many in England.   They
also, when the waters are clear, yield
trout in the manner which now and then
affords an engaging variety to fishermen
in the south-west of England; but sometimes they are in a state which calls for
another method of angling.   That is the
state of flood.   The local anglers are
always hoping for it, and they hail its
coming with delight.

One can share their feelings.

For weeks the stream has been steadily
falling; there is in it so little water that
the millers all along the course have had
to push down the sluices of the dams o'
nights, so as to accumulate force for use
in daytime; the river is but a shadow of

itself, much too small for its bed. Then, in June let us say, there are signs of a change in the weather. The sportsmen in the village become alert, and vie with one another in prophecies of flood. The mercury in the blacksmith's glass has been creeping down for days; shepherds come in to the weekly market and report signs of storm in the behaviour of their flocks; the veterinary surgeon, who is all over the county, says that when driving home late last night he saw sheet lightning on the southern uplands, towards the sea.

Into their gardens all the villagers go to dig for worms, and soon each has a few hundred snugly bestowed in a bag of moss damped with cream. The sun goes down behind long banks of motionless thick clouds; but, alas, the rain holds off. Next morning the earth is still dry; but all the sky is gray, and the ancient weathercocks, which are rather rusty and not responsive to trifling airs, show that during the night there has been a

considerable puff from the south-east.
Ah, that is better! It comes! There is
actually rain at last! So strange is it
after the long drought, the villagers go
forth, hatless, to make sure. They cannot but believe it when they feel it.   It
seems in earnest, too : not a violent
burst that passes as sharply as it comes,
but a deliberate slant of small drops,
which, if they were frozen and the time
was winter, were heralds of a feeding
storm that in a round of the clock
would wrap the country thick in snow.
On the rain comes, increasingly ; it is
noted with joy that it does not pause at
twelve o'clock, which would mean a risk
of its stopping altogether ; and by five in
the afternoon there is no longer any room
for fear.   Certainly the floods are out!

Some who have been down to look at
the stream announce that there is no
change there yet ; but that was to be expected.   The ground has been very dry :
it has to be thoroughly soaked before
the water begins to run.   Besides, the

wheat and the oats and the barley, the turnips and the potatoes, have to be served before the stream. If it were autumn, and the fields were all stubble or fallow, the river would have risen a foot by this time; but the "growing crops" drink up a large quantity of rain.

All is still well at eight o'clock. News comes that the burn which runs for a few miles by the side of the North Road, and so drains a good strip of bare land, is rising so quickly that the river, below where the burn joins it, is muddy for two or three yards out. Some of the larger ditches are beginning to run.

Meanwhile the rain goes on: no longer a slight windy spray, but coming steadily down through motionless air, pattering on the leafy trees: the freshened earth is alive and awake, purring in gratification.

Suddenly there are twitterings in the gardens, and the copses ring with the notes of thrush and blackbird. That makes the villagers uneasy. The birds sing when the rain is past: is it about to stop? Happily, the woodland music, which was over in a few minutes, seems to have been a false alarm. The rain is better than ever. Water is gurgling down the eaves of the cottages, trickling over the pebble paths in the gardens, and racing in the ditches beside the high road.

It is now nearly ten o'clock, and the eager villagers go to bed.

They are not there long. You are an early riser indeed if you are first on the stream in the morning. Rather is it likely that every fifty yards or so you will see a villager, rod in hand, the point of it low down near the surface of the stream while the end of the butt supports his elbow, moving very slowly along the bank. All the fishermen in the little community have been out since break of day. Intently watching his line, which, you notice, is very close inshore, each is moving with the bait as the current bears it down.

Perhaps there has not been much sport thus far. Indeed, it is probable that there has been none. The trout are not in good humour at the first flush of a flood. Then the water is very thick, full of the waste matter that has been accumulating in the drains and the ditches, and on the roads, for many weeks; and perhaps the fish, though not easily disturbed, are off their food.

Soon, however, all comes right. The stream, which on first being affected by the rain was of a rather noisome mixture of various grays, gradually becomes clay-coloured, with a tinge of ruddiness; by and by, as the flood begins to fall, it will be a delicate yellow. It changes so, not wholly in response to the changing lights, though the buoyant clouds, which, as is usual after a summer storm of rain, are scurrying from the north-west, are of varying hues; but mainly from material conditions in the stream itself.

When it assumes the yellow tint, the trout begin to bite; and if the weather

keeps up they bite with a will all day.
What a peculiarly agreeable day it is!
Often I have wondered what it is that
causes such a time to be remembered, or
looked for, with such pleasure.   It is a
rare day, no doubt, floods in late spring
and in summer being infrequent; and
perhaps the joy with which one contem-
plates the sport is in some measure due
to its novelty.   Still, that cannot be the
whole explanation.   Angling in a flood
has an attractiveness inherently its own.

After much pondering, I have, I think,
hit upon the secret.

A rise at a fly, delightful as it is and
always will be, is the joyous sensation of
a moment only; but a nibble at a worm
is more.   It is a protracted sensation.
If you watch any of these villagers who
are out upon the stream when it is flowing from bank to brae, you will notice
that he does not strike the very moment
his line is stopped.   O no: this art of
worm-fishing calls for much discrimina-
tion.   It may be an eel that is taking

his bait. If he could be sure, he would instantly pull the line away, not wishing to catch an eel; but he cannot be instantly sure. The eel's nibble causes a slow and lazy-looking movement of the line, and the trout's is usually a smart rug-tug-tug-dart; but often eel and trout begin in the same way, which is merely by arresting the line. The angler must risk catching an eel to make sure of not losing a trout. On the other hand, he must not wait very long. When a trout finds that he has made a mistake, he has an unknown means of putting things right which is nothing less than marvellous. Sometimes he ejects a hook as neatly as the mechanism of a modern rifle ejects the shell of a cartridge; often, if he fails in the attempt to do that, you will find, on taking him out of the landing net, that in doing his best he has at least detached the worm from all the three hooks on your Stewart Tackle and blown it a foot up the gut! The trout must have a strength as magical as that of the mole, which, for its size, is

said to be the mightiest of animals.   All
this time, nigh half a minute, our fisherman has been watching the line.   In the
question of when to strike a very complex tangle of considerations is involved.
What he is to strike is the most serious
of all.   It may be an eel; it may be a
trout not larger than a herring; as fish of
all sizes are susceptible to the attractions
of a worm, it may be the monarch of the
stream.

Does not this explain the delight of
the time when the summer floods are
out ?   Frequently during the day all the
pleasurable excitements possible in the
sport are wrapt in a few tense seconds
that feel much more.   Indeed, recollected
in long retrospect, the joys of a good day
in a June flood seem almost to cover a
season.

It is natural to expect that when the
stream has cleared fly-fishing will be
much better than it was before the rain ;
but this hope is not always justified by
the event.   In lakes a rise of water

almost invariably brings an improvement in the sport of fly-fishing; but it does not seem to make much difference on rivers. Often, on the fall of the flood, as the sparkling water was running half a foot or so above the normal level, I have thought, Now, this is splendid; but nearly as often I have found it not splendid at all. The water is clear enough to let the flies be seen, and it is flowing with such liveliness that one would think the trout must be lively too; but often that is not the case. As a rule, the fish hold aloof until the stream has become normal.

Then a river affords opportunities to study the habits of the trout which are not to be found on a lake. Many of the fish can be seen and watched. Does each of them have a place of his own? I think he has. Sometimes you may see a trout, usually a large one, roaming about within a radius of a few yards; but when you see this you see also that there is no other trout near him. All the

little domain in which he moves is his;
and if we watched long enough we
should probably find that when he rests,
or feeds on flies, he is stationary at some
particular part of it. Here and there,
most notably where a tributary joins the
stream, three or four trout are often to be
seen together. These hardly ever move
from the spots on which they are lying,
or above which, as they will be if on the
outlook for flies, they are poised. Each
seems to think that if he went away for a
while he would have a battle for his place
on coming back. These three or four
trout, too, are in a distinct order of precedence. The biggest is closest to the
point of contact between the tributary
and the stream. Being there, he has first
choice of the tid-bits which the brook
or the ditch is bringing down. Next to
him is the second-biggest; next again,
the third-; and so on. When one is
taken by an angler, his immediate junior
has a step in promotion. If all of them
are taken, next day three or four more,

of equal sizes or nearly so, will be found
in their places.   Whence they have
come, no man can exactly tell; but there
they are, mysteriously, and it is reasonable to believe that they had been
looking for the vacancies which they
have filled.

Once I saw this little drama complete
in a single day.   That was on T——
J—— B——'s water rising on the borders
of Surrey and Sussex and Hampshire.
The two ponds near the old mill are
separated by a grass-covered path, across
which the stream, having filled the upper
pond, falls into the other.   The second
pond is about nine feet under the surface
of the first; and the stream tumbles into
it perpendicularly.

Now, just below the waterfall lay
three great trout.   They were well
known.   Approaching carefully, any one
could see them from the grassy path
above.   They had been there, the second-
largest just behind the first, and the third-
largest just behind the second, for years;

they were famous in the hamlet, and had
been heard of far beyond; rumours of
them, indeed, had reached Godalming
and Hindhead, and even Winchester.

Well, the first day he took me to that
stream, T. J. B. showed these fine fish
to me, and dared me to catch them if I
could. "It would be rather a pity, of
course," he said: "I regard them as
privileged retainers, you know: but," he
went on, pleasantly, "you may try your
hand at them. Every one else who has
been here has tried. I even brought old
Farley, the gamekeeper at the shooting in
Kent, who's very good with the fly-rod,
to cast for them; and he couldn't manage
it. Farley said he might get one of the
trout if he could throw into the waterfall
from the opposite side of the pond, which
would be fifty yards, and so was out of
the question, but that there was no
other way of getting at them. The
bushes, you see, come out from the bank
at both sides of the waterfall. Do you
think that from this side of the pond, on

either side of the fall, you could flick a
fly in round the corner of the bushes, and
so drop it over 'em—some modification
of the Spey-cast trick ?   If so, go and
do it ! "

I went to try ; but soon perceived, as
I had expected, that to make the fly
break in round the bushes was much
more difficult than slicing or pulling at
golf, or at cricket making the ball break
in to the middle stump.   In fact, I inwardly agreed with Farley that it could
not be done, and that the three great
trout were likely to remain there as long
as the mill dams stood.

The mind, however, has obscure ways
at times.   Occasionally, it would seem,
it is at work on its own account, and
reveals the fact that it has not been idle
by suddenly presenting a solution of
some problem that had apparently been
abandoned.

About   two   hours   afterwards,   at
luncheon a good bit away from the ponds,
I realised that I had exclaimed, " I've got

it!" "Got what?" asked B., who was distributing among his guests flagons of claret and nut-brown ale. "That great trout, and perhaps the others too!" B. laughed; asked what the idea was; and offered three to one, in new hats, against it, whatever it might be. I took the bet, adding that the hats would have to be of different kinds. "All right!" said B. "Are you to catch them with a fly, or what?" "Yes: a fly; but it is you who are to catch them—I am to tell you how."

After a brief rest amid the fragrance of pipes and wild-flowers, B. and I went back to the ponds. I examined his cast, and took off the upper fly and the middle fly, and saw that the remaining one, a Red Palmer, was sound. Bidding my friend keep well out of sight, down I lay prone on the grassy path by the side of the waterfall, and peered over. All the three great trout were there. "Now, B., give off two or three feet of line, no more; put the fly into the middle of the trickle between the bushes, just before you, and let it fall with the spray. When I whistle, strike—but gently!"

Down dropped the fly, daintily touching the water of the pool a foot in front of the first trout, and a foot to the right of him, and then sinking. The fish turned and looked at it, but let it pass; and the trout behind him took no notice. That was discouraging. "Again, B.," I whispered: "just as before." This time, the moment the fly fell into the pool the trout came up a few inches, turned, without touching the water, and sank quickly back. I whistled low; and before I could scramble to my feet B. was tearing along the grassy path at the wrathful tail of Number One.

Round and round the pond he had to scamper, and round again, before the trout could be cajoled or coerced into the landing-net. The fish weighed six pounds and a half.

The others were caught in the same manner before it was time for a cup of tea at the inn close by. Number Two weighed five pounds; Number Three, three pounds and three-quarters.

When the carriage came to take us back to H—— for the night, I peered over the waterfall, and saw three other trout exactly like those the capture of which had left our host in a state of high exultation. I think that the place they occupied is a favourite in summer simply because it is there that aerated water plashes into the pond.

I have stated a belief that when a trout has chosen a position in the stream he stays there; but that, of course, is only during the three months or so when the season is at its height. Before that period, in early spring, the fish undoubtedly move; but they move in a regular way. During the winter they have been up the tributaries, or in the shallows of the higher reaches of the main stream, spawning. After that they drop down and rest awhile in the slowly-moving deeps. Late in March they begin to appear in the rapids. By the

end of April they are in the places that
were probably their haunts the season
before. There, if not caught, they
remain until the first early-autumn flood,
on the coming of which they begin to
move up-stream. If the flood is considerable they congregate at the mouths
of tributaries. They are extremely
voracious at that time. Many of them,
when caught, are found to be filled to
the lips with worms and grubs and flies.
Their hunger, I think, is due to the
demands made upon them by the rapidly-
ripening eggs and milt. Soon after that,
as is proper, they pass into the care of
the gamekeepers, and the angler has
for a few months sheathed his rod.

One cannot so closely observe the
habits of trout in the great waters that
are still. This lends a pleasant mystery
to the lake. The pleasure would be
abated if the mystery were solved or
lessened; yet, such is the perversity of
man, I have been constantly trying to
solve it.

Day after day in early spring, as has
been mentioned in the chapter on Tem-
pérature, the sport was disappointing; and
that excited the thirst for knowledge.
How many trout were in the loch?
Were a large proportion of them very
big? Did they all, like fish in a river,
lie with their heads in one direction? or
did some of them look one way, some
another, some another, and some another
still?

Yes, my host said: the trout in the
loch were very plentiful, and many of
them were very big. As to how they
lay, he was not sure; but his impression
was that in a breeze they always kept
head - to - the - wind. These things, he
added, I could easily find out for myself.
All I had to do was to go up the hill at
noon some calm day when the sky was
clear, and look down upon the loch
through a powerful telescope. I should
then have a wondrous spectacle. Wherever I looked I should see the trout,
closer to one another than grouse in a

covey, poised about a foot under the surface, watching the insects, and rising at them now and then. I should see uncountable thousands of the fish; and there would be other thousands far below, large ones that rarely took a fly but were always ready for a minnow.

Up the hill I went about four hundred feet, and, preparing for the survey, seated myself on a boulder.

It was a fine morning. In the motionless air, the valley was flooded with soft spring sunshine, dead - still upon the heather, which bore the russet hue of winter, and slightly shimmering on the tender green of early-budding trees; and the narrow loch, dark-blue, was like a mirror. It was almost difficult, as one gazed, to be sure where the land merged into the water. After a few moments' looking at it, the long sheet, being quite still, lost the aspect of water: it seemed to have vanished, and the space which it had occupied to be flanked by mountains of giant stateliness and repose: only when one shook oneself to break the spell was there any sense of incongruity in beholding hills tapering from their bases downward.

All the conditions favoured the purpose with which I had climbed ; but what was this ?

Below, a little to the east, rather more than a gunshot off, something was happening.   In the midst of the sunlight, floating a little above the hillside, was an unshapen column, too fragile to be thought material.   It did not intercept the beams.   Permeated, indeed, it was with these, which seemed to be all the brighter for having something to play upon ; and there was no shadow on the ground beneath.   The column moved. Almost imperceptibly, it was changing its shape at every moment.   Very slowly it was coming upwards, and was growing larger.   On it came ; on, and on, and on ; silently ; radiant and softly sparkling, and that not only on the irregular edges, but all through its apparently impalpable mass; it was like a scene from fairyland in the broad light of day. By and by, without having noticed the contact at the first moment, I found myself swathed. The column was around me, and above; quite high above, I noticed on looking upward; around and above, too, the strange column, or rather that of which it was composed, small fluttering white things, still caught the sunbeams, and seemed to toss them to and fro.

It was snowing; and while the small fleecy crystals fell, as is their wont, the exquisite thin cloud from which they came, the magical column, rose!

In a few minutes, as mysteriously as it had appeared, it ceased to be; and where it had been, the tranquil sunlight lay.

This made one think about certain speculations of the learned. These flakes of snow, beautiful, each in perfect harmony with a design beyond all human ken, could not, in one sense, be said to have been evolved. They had

come instantly, each in a flash, although
the column of them had been slow in
motion and in growth; and they had
not come out of anything that the eye
of man could see. Of course, there was
the air, and in the air was moisture; and
it was of air and moisture that the flakes
had been made. What made them?
Perhaps the making of them had begun
millions of centuries ago, when the fluid
mass of nebulæ that had solidified into
our globe had been sent spinning on the
course ordained. To that their approxi-
mate origin really might be traced: it was
conceivable that what I had witnessed,
the action of a wandering chill, was the
inevitable outcome of forces that had
been set in motion countless æons before
man appeared upon the earth, before,
indeed, the earth had a separate being.
Yes: the veriest Calvinist must see some
cogency in the theory of Evolution,
especially when it is applied not to the
species to which he himself belongs,
but to the species of equally marvellous

things that are inanimate. On the other hand, was not all this a playing with words, or, rather, a playing with what words represent, those ineradicable necessities of thought which spring from the impact of phenomena upon the reflective consciousness? Although they are ineradicable, they are not necessarily right. If mountains stood on their heads, and trees grew with their roots in the air, and birds walked the earth while wingless animals flew, and trout rose at flies through a yard of ice, it would be all the same: some argument into design, and from it, the familiar system, would be sure to be advanced in explanation of these circumstances, just as it is advanced in explanation of things as they actually are. Things must always be somehow, and, however they were, the human mind would strive to interpret them, and think it did : a sense of need for synthesis is inseparable from the human understanding. Synthesis is of various conceptions, however ; and there's the rub.

Did the First Cause yield up the power
of causation when the whirl of the con-
stellations was set going ?   Calvinist and
Evolutionist assume so ; but, somehow,
one doesn't quite like the tone of either.
They take so much for granted that
their dogmas should be small, and even
then should have less of the last-word air
than they habitually carry.   If the First
Cause did not surrender the faculty of
causation ?   If it be retained to this day ?
That might account for the fairy snowstorm.   It was just as wonderful a thing
as the earliest appearance of any species
of animal can possibly have been.   Out of
the dust was man made ; out of moisture
and air, the snow.   Man reproduces the
species, and the snowflakes don't : one
is a specific creation once for all ; and
the other are a special creation, constantly in repetition by forces outside
itself.   That is a difference, certainly ;
but is it as wide as it seems ?   May not
the coming of each new human creature
be in a sense just as special an act as that

of each new shower of snow? Is the
perception that the First Cause ordained
the species incompatible with belief that
the First Cause may still control the
creation and the varieties of the units?
Can the occasional coming of a genius
such as Shakspeare be accounted for by
"natural laws" undisturbedly arranging
their own results? That is a thought
apart. The immediate question is,
Were these wonderful snowflakes the
result of a fiat issued that very day, or
were they the result of something that had
happened millions of centuries before?
On the whole, it seemed just as likely
that I had witnessed an instantaneous act
of creation as that the flakes were merely
the inevitable outcome of the state of
things which was arranged just after the
earth ceased to be without form and void.
You see, neither Calvinist nor Evolutionist, though both dogmatise from perceptions that one must admit to be
necessities of thought, is at all times conscious of all the necessities: some of

the perceptions exclude others that are
equally to be respected. Is not this a
notion quite clear, irresistible, absolutely
impossible to escape from, when one
mentions it: If the First Cause surren-
dered the faculty of causation immediately after the original creative act, there
ceased to be an Omnipresent Cause at all,
and God created the world only to leave
it godless? Thus does it seem that
Calvinist and Evolutionist are head-
masters in schools of dogmatic Atheism,
one quite as questionable as the other;
and that .

At the hospitable place where I was
staying, when any one is on the hill or
far out upon the lake, it is a horn, instead
of a gong, that proclaims the approach
of a meal. A blast rang through the
valley, and shrilled off echoing in the
corries. On rising to go down, I saw
that a catspaw breeze was creeping
over the water, and that the pleasant
mysteries of the lake must for a time
remain unsolved.

CHAPTER VIII

THE " WHUSTLER "

Waiting for the Wind—An Unexpected Rise—" A
Birr! a Whirr! a Salmon's On! a Goodly Fish! a
Thumper!"—Involuntary Cruisings—An Alarm-
ing Dive—The Salmon Sulks—A Stirring Squall
—Ronald in Despair—Cast on a Strange Shore
—No Gaff!—The Outflowing River—A Disquiet-
ing Prospect—Pull-Devil, Pull-Baker—Honours
Even.

WHEN Ronald and I set out on Loch
Voil the weather was unusually promising. In the morning there had been
squalls charged with rain ; but now, just
after luncheon, the wind was steady.
Surveying the hillsides of the glen in
which the water lies, one could now
and then see a patch of heather or of
bracken gently gleaming in sunshine.

That showed the clouds to be thin and
airy At length, apparently, we were
to have a good day. Anglers will know
what that means. Others will regard it
as an unimportant remark, and will perhaps say that fishermen, like farmers, are
always grumbling. Those who are neither
fishermen nor farmers are strangely
ignorant about the weather. The out-
standing facts are plain to them; but
they are not conscious of the grada-
tions and other subtleties. They know
when there's rain, or heat, or cold,
or a gale; but when they go forth to
business of a morning feeling chilled a
little they say, "Ah! an east wind again,"
although probably it is from the west,
and are unaware that the force of the
wind varies from minute to minute. The
knowledge which they lack is possessed
by anglers; and that is why, having a
strange story to tell, I begin about the
weather. It is all-important. If the
wind is strong the boat drifts so quickly
that in playing one trout you pass over

places in which others might be expected.
If it is of the fitful, gusty kind that
sometimes comes when there's thunder
lurking about, the fish are sulky and
don't rise.   If there is no wind at all,
what are you to do ?   The boat won't
move unless you pull it.

The last-mentioned predicament befell
Ronald and me.   We had not been five
minutes afloat before our soft breeze
drooped and died.   We had intended to
go to the head of the loch, where there is
a large sand-and-pebble shallow, just
the place where sport is to be hoped
for in a good wind ; but, now that the
breeze had passed, there was no use
going.   Indeed, was it any use going
anywhere ?   I put it to Ronald frankly,
but with chagrin.

"'Deed, ay, sir !" said the gamekeeper
reassuringly.   "Ye have to throw the
flees lichtly in a dead calm like this ; but
if ye manage that ye often raise a troot."

This I knew.   In a smooth stream a
dead calm does not put a stop to one's

sport: why should it render hopeless fishing on a lake? Only because the flies and the gut which one uses on a lake are as a rule heavier than those which one uses on a stream. The cast I had on was not at all a thin one; it was stout enough, indeed, to hold as big a trout as could be expected; still, there would be no harm in trying. Perhaps the wind would be back ere long.

Out on the deep, then, Ronald slowly rowed, and I kept casting as we went along. Not a trout moved. The water was so still that the scenery was reflected on it with bewitching minuteness of detail. As you gazed steadfastly, there seemed to be no water at all, but only space, with two ranges of hills converging downwards, downwards, until, very far down indeed, they were standing on their snow-capped heads. It was a spectacle the paradoxical fascination of which made one giddy.

"There's a rise, sir," said Ronald: "wull I pu' to 't?"

It was a relief thus to be recalled from looking upon the Highlands upside-down. We pulled towards the rise, the expanding ring of which lingered on the water; but, although the flies fell lightly over where the trout was, the trout remained below. So it was with a good many other trials. Like hunting the fugitive ripple when the air is faint, stalking the rising fish is sometimes a fruitful occupation; but it was of no use that particular afternoon.

Ere long we reached the head of the loch. "Wull we try Doine noo?" Ronald asked. Lying to the west, Loch Doine is connected with Loch Voil by a short, deep, slowly-moving river. I was not sure whether it would be well to go into Doine. If the wind, when it rose again, should be from the east, we should be favourably situated as regards Doine, having only to slip through the river, with a drift the whole length of the loch before us. On the other hand, if the breeze should come from the west,

we should be equally well-placed on
Voil. So I answered :

" Let's wait a little, and see where the
breeze is to come from. It will probably
be either from the east or from the
west."

" Ay : that's so," said the gillie.
" There's never a north or a sooth wind
on they lochs. The cloud-carry may be
frae ane o' they airts ; but the hills block
the wind, and it aye soops up or doon the
glen."

I laid aside the rod, and prepared to
smoke.

" That's a dainty bit wand," said
Ronald, taking up the rod and making
a gingerly cast. " Nae mair than nine
feet long, I'se warrant ; and as licht as a
heron's feather."

" Only five ounces, without the reel,"
I answered proudly. " It is a present
from America. Built-cane, you see, and
quite strong—the friend who gave it to
me says there's not a trout fit to break
it in this over-rated island."

"No?" said Ronald, who during this brief dialogue had been testing the casting power of the little rod.—"Guidsake, what's that?"

It was for him, rather than for me, to say; although out of the corner of an eye, as I was screening with my hands the flame of a match, I saw a disturbance just where the flies had fallen. It was a sudden surge in the water and a furrow heaving outwards.

"She's a whustler, whatever," said Ronald eagerly. "Tak' the rod, sir?"

"No, no, Ronald: your bird, you know. Does he feel heavy?"

"Vera," said he in quiet wonderment. "A whustler beyond a doobt."

"Whustler" means big and fierce fish, probably so-called from the peculiarly agreeable tune which the reel plays as the line is run off. Thus, Ronald's statement was very cheering.

"Michty me, look at that! Tak' the rod, sir—tak' the rod! We'll ha'e to pu' oot."

"That" was a large dorsal fin and half
of a majestic tail angrily protruding, and
then a long dark-blue back, as the
whustler, now thirty yards off, cleft his
way.

Ronald handed me the rod imperiously,
and sat down to the oars, pushing outwards stern-first.   There were about forty
yards of line left on the reel, and these I
was yielding foot by foot.   Ronald's most
vigorous efforts with the back-watering
oars were scarcely sufficient to prevent
disaster.   If I paid out no line at all,
something would break; if I let it go
freely I should soon, with the same
result, be at the end of the tether.   My
legs began to tremble : they did not seem
to be based on anything substantial.
Still, I contrived to speak with astonishing composure :

"What's to be done, Ronald ?"

"Am thinkin', sir, ye'll better step
over to the bow.   Then I'll turn the
boat, and be able to follow her faster.
Canny, canny !" he added, as I stumbled

16

across the thwarts. "If ye let her slack
a second she'll slip off, and if ye're too
tight she'll break ye!"

Thus admonished, I found myself
standing with dignity at the prow, gazing
out on the mysterious deep, somewhere
in which the whustler was still unmistak-
ably on. He showed as yet no violent
excitement : only, away he went, steadily,
unrelentingly, the boat in pursuit as
quickly as Ronald could drive it. Within ten minutes we were in the middle of
the loch, which is much less broad than
long. Suddenly the strain yielded. To
my horror, I found that I could reel in
without resistance. Sick at heart, I turned
and looked at Ronald. He was rowing
with might and main.

"Stop, Ronald."

He looked at me, over his shoulder,
in apprehensive interrogation : clearly he
meant, "Is she off?"

"I think so," said I; and was beginning
to assure him that I had really made no
mistake, when the sound of a heavy splash

just behind caused me to wheel round to attention at the prow once more. To the left, not more than ten yards off, was a circle of writhing water.

"I saw her,' Ronald was exclaiming in low tones; "and she's no' off yet. Reel up, sir; reel up like the tevil when ye've got the chance."

Obeying, in less than a minute I had the happiness of discovering that Ronald was right. The whustler was not off. He had merely changed his tactics. Perhaps he had leapt to snap the line; perhaps——

This was no time for conjectures. The fish was running down the loch at a very rapid pace. Like a living thing on lightsome wing, the boat sped before the oars as it never sped before; yet the reel was screeching. Just as the end-of-the-tether crisis was at hand, the whustler slowed down a little: indeed, it was possible to recover a few yards of line.

"That's richt, sir,' said Ronald encouragingly, but rowing as hard as ever.

"Aye reel up when ye can. It pits off the evil hour."

The evil hour! At times of excitement the imagination is alert, active; and Ronald's words started a new train of thought. When was the evil hour to come? Already it seemed a long time since the whustler had made his presence felt. Already we had gone careering after him through the little bay lying to the south of the river from Loch Doine; thence we had crossed the mouth of the Monachyle Burn: these were landmarks on the northward course. On the way down the loch, Monachyle Mohr was already far behind; we were now flying past Rhuveag, a pretty cottage from whose chimneys the blue smoke of wood-fires was lingering opalescent among the dark-green pines in the background; soon we should be at Craigruie Point, off which the loch is unnavigable when the west winds are out in earnest. The evil hour! Were not we in pretty evil case already?

Ronald himself seemed to think so.

"This," he said, "looks like a long job. She'll no' tire for a while. Ye needna' gi'e her the butt—the bit wand would just bend and she wudna' feel it. Am no' muckle in favour o' they new-fangled split - cane toys. Gi'e me an auld - fashioned greenheart — something ye can hud on by. That fish micht vera near as weel be free a'thegither. It's no' us that's caught her — it's her that's nabbit us."

This seemed true. As far as I could make out, we were no nearer capturing the whustler than we had been before he took the fly. He was not now tearing through the water quite so fiercely ; but I had no confidence that he was without reserve of strength. Certainly he was full of resource. He had turned to the right, as if to pay a call at Muirlagan Bay, and was apparently wagging his head from side to side. I felt that the gut might give way to one of his uncomfortable tugs.

"What do you think he is, Ronald?
A big trout?"

"Na."

"A ferox?"

"There's nae ferox here. This is a
weel-bred loch."

"A salmon, then?"

"A salmon sure enough, sir; and a
thirty-pounder unless am much mista'en.
I saw her loupin' when ye turned roond
thinkin' she was off."

"But what did she take the fly for,
Ronald? Salmon don't feed in fresh
water—so they say nowadays."

"That's a' damisht nonsense. What
for should they starve in fresh. water,
sir? Because ye never find flees or meen-
nows or onything else in their mouths,
or inside them, when ye catch ane? As
weel say that they dinna' feed in the sea
either, for the same reason; and that,
thairfore, they pit on four or six pounds
weight every year on naething at a'.
Whaur's she off tae noo?"

The whustler had again changed his

course, and was making for Ledcriech, on
the north shore.   We followed submis-
sively.   Ledcriech Bay is made beautiful
in summer by water-lilies.   These were
not in blossom just then, so early in the
year; but I dared say that below the
surface the stalks were in tough abundance.   What if the fish got in among
them?   Could we ever get him out?   I
had misgivings; but I did not like to
mention them.   Ronald was not in the
best of tempers.   He seemed to think
that we were having an untoward afternoon, and that I was responsible.   Among
other misfortunes, we had no gaff aboard.
I felt that he was thinking of this, and
assuring himself that it added to the
certainty of the evil hour.

Fortunately, we did not reach the
water-lily bay.   A considerable time
before he could be in sight of the opportunity offered by its harbourage, the fish
was cruising down the middle of the loch.
It was not at all easy to keep up with
him.   If I could have spared any

sympathy from myself, I should have bestowed it upon Ronald. Although the sun was now sinking behind the western peaks and the evening chill had come, Ronald was sweating, and, not having foreseen the possibility of this how-d'ye-do, we had set out unprovided with the means of refreshment.

The tension changed. Instead of keeping on the forward path, the whustler seemed to go straight down. Down, down, down he bored, getting leave of the line only because the boat, although Ronald was stopping her, was still going towards the place from which the dive had begun. Down, down, down : when we were practically straight over him he was still diving, taking the line from the reel. Here was a new peril. About this place Loch Voil is at its deepest. If I remembered the chart rightly, the depth was very great indeed. Would the line of the little trout-rod suffice? If not, should I supplement it by dipping down rod and arm on the desperate chance that

the extra twelve feet thus gained would
be enough ? At the moment I had no
thought for the ludicrousness of the pros-
pective situation. Humour flees from
fright.

Much to my relief, the line itself
sufficed, and there was even a little to
spare. Whether the salmon had gone
quite to the bottom or not I cannot say ;
but, wherever he was, he stopped. He
moved neither to right nor to left, neither
up nor down ; but he was still on. Of
that there was no doubt. I had never
lost touch with him during the dive ;
and I felt him still, though he was stead-
fast ; and through the line there ran a
tense quivering thrill like that of a
telegraph wire. The little rod was
trembling as my legs had been at the
beginning of the episode. Being now
well inured to the crisis, I myself was
comparatively at ease.

So, I noticed gladly, was Ronald, resting
on his oars after nigh three hours of hard
and anxious toil. Five minutes passed ;

ten; fifteen; and then it dawned upon me that, though tearing over the loch at the truculent will of the whustler had been fearsome work, we were not now very much better off. At least, we were not perceptibly further forward. There was no disguising the fact that the enemy had us at a disadvantage. Excepting that I had to keep in constant touch with him and be sure he was still there, we had nothing whatever to do. The shades of night were falling; we were fixed on a cold wilderness of water with neither food nor drink; and it had become evident that we might have to stay there indefinitely unless we were willing to cut the painter and scuttle home defeated and disgraced.

That, of course, was not to be thought of.

" What's to be done, Ronald ? "

" That I canna' tell, sir. I've never been in sic a scrape as this before."

" O, surely: it often happens: a salmon often lies doggo."

" Never like this that I've seen; though it's true enough that, exceptin' when I went to the war wi' Lovat's Scouts, I've never been anywhaur else but Glenartney Forest and here."

" I've seen it happen on the Dee."

" Ay; but the Dee's a river, no' a loch."

" On the Dee, when a salmon lies long at the bottom of a pool, the gillie can always get at him and stir him up some-how."

" Nae doot; but the Dee's no' scores o' fathoms deep."

" The gillie sometimes throws big stones at him."

" In this boat there are nae stanes, either big or sma'."

Ronald, with his cold logic, had undoubtedly the best of the argument, which, indeed, I had initiated less from having anything to say than from a vacuous feeling that silence would seem a confession of helplessness. It was true that I had seen a gillie stoning, and thereby putting to flight, a sulking salmon in the Dee, at Banchory; but I had realised, even as I mentioned this, that such an expedient was out of the question on Loch Voil. It is astonishing how a man chatters when in a dilemma. Contemptuously irritated at myself, I turned upon the gillie in wrath and mixed metaphors.

"Chuck it, Ronald," I adjured him. "What's the good of sitting there wise as an owl and depressing as a wet blanket? Buck up. We've got to kill this salmon."

"Ha'e we, sir? There's mony a thing we've got to do that we never do."

"Come, come, Ronald. That's no talk for a Lovat Scout."

Ronald was not pleased; but he answered reasonably:

"That wark was naethin' to this, sir. In the war we aye kent that onything was possible, and did it; but in fishin' some things are clean impossible, and this is ane of them. She was a cunnin' man, the Boer; but she was an innocent babe to this fish."

"Dry your eyes, Ronald. He'll surrender some time."

"No' she. Ye dinna' seem to understand, sir. D'ye no' see that when she starts again after this long rest she will be quite restorit—just as bad as if we had never run her at a'? Wi' that wee toy o' a rod, ye've dune her no harm whatever. If we ever get oot o' this, and ha'e to dance after her again, it will just be as if you had hookit a new salmon, and we'll ha'e the same business a' ower. I see nae end tilt."

Neither did I; but I saw something else. Although the light had almost gone, I saw that there was a ripple on the water at the head of the loch, far away. It was coming towards us rapidly. Soon, too, the sound of the burns on the hillsides began to grow in volume and in briskness. Hitherto the noises of their falling waters had been soft and hushed, half lost in the immediate still atmosphere absorbing them; but now they were loud, and growing louder, almost harsh. That meant the coming of a wind. Would the wind awake the whustler? Time would tell. It did; and soon.

When the curl on the water reached us Ronald took to the oars again. A very slight breeze is sufficient to set a boat moving; and, of course, the extent of our line allowing next to no latitude, we had to keep, in relation to the whustler until he moved, nearly perpendicular. That was not a task so easy as those who are unused to boats may imagine, and Ronald did not enjoy it. Each minute the air, at first a zephyr, was increasing; and amid such conditions it is almost impossible to keep a boat exactly where you want. A few yards in any direction would again take us to the end of the tether; and then?

Happily, the need to consider the query was postponed. The whustler moved. Perhaps the ripple attracted him. The surmise was in accord with a theory which I had been cherishing in secret, and for a moment I thought of broaching the argument to Ronald. A discontented gillie, however, is not an appreciative audience for speculative thought; and I held my peace on all save the topic of the hour.

"Well, we're off again," said I, cheerily, hoping to quiz Ronald out of the doldrums.

"Quite so," he answered; "and practically, sir,—practically, mind ye—it's a new salmon we ha'e to deal wi'—just as fresh and ferocious as if she had only this minute risen at the flee." To himself he added, muttering, "And a bonnie time o' nicht to begin the day's sport!"

I could not understand Ronald. As a rule he was the best of gillies, grudging neither time nor trouble in the pursuit of game, keen and joyous as Tim the terrier in a rabbit warren. There are bonnie lasses in Balquhidder; and Ronald is a youthful warworn hero; and perhaps

Spring, which, it will be remembered, deals in a livelier iris,——

"Steady, sir, steady! Sit doon!" exclaimed Ronald, interrupting my apologetic reflections. "See yon!" He nodded westward. I turned for a moment to look.

To within a hundred yards of us, all the loch was churned and seething white, and the dark air was gray with sleet.

Having had some little experience of the storms which suddenly descend upon Highland lochs, I did not like the look of things. Indeed, inwardly I began to sympathise with Ronald's view that we should have anticipated the evil hour by cutting ourselves free from the whustler long before. However, the time was not suited to after-thoughts; and I pretended not to understand.

"Right O, Ronald! The gut, I think, will hold—sound Lochleven."

Meanwhile the whustler had led us a considerable distance from the place in which he had rested and been refreshed. As it was now impossible to see the

shore, or even the point of the rod, I
could not say how far we had gone ; but
I felt in a general manner that we were
still on the eastward course. Ploughing
industriously on, the fish had been
making no undignified display of anger :
indeed, I had come to regard him with
the familiar affection in which one holds
a good retriever, saying to him, as occasion
required, "Steady, lass!" or "To heel,
you devil!" or other caressing phrases
of the field ; but with the progress of the
storm our relations became strained. He
began to leap. We could not see him ;
but we could hear him well enough amid
the short thick thuds of the waves beating
on the boat and the baritone boom of the
squall. It was, I confess, an alarming
sound. At each leap I expected the
performance to be my last. That seems
a strange remark ; but it is accurate.
When he was down in the water and
could be felt, I was not without hope ;
but that was momentary only. Whenever the line slackened I knew he was

17

aloft in the air, and my heart stopped. Ronald was in similar extremity. The salmon seemed to be aimless in his movements. At any rate, his leap was sometimes on one side of our creaking craft, sometimes on the other; now off the stern, anon off the bow. Thus, Ronald was in perplexity. Sometimes he had to pull away from the fish; sometimes to push towards him. All through this trying time the general drift of things was determined by the wind, which we believed to be still from the west.

"This canna' go on much longer, am thinkin'," said Ronald. "I daurna' pu' either to the north shore or to the sooth, for then we'd be broadside-on and be blawn ower. Forbye, the boat has been lyin' up a' winter, and is brittle. If ane o' they big waves catches her on the side when we're turned to follow the fish, she'll be staved in. I doobt we're by wi't, sir."

Although he had to shout in order to be heard, Ronald delivered this grave opinion in a deliberate, matter-of-fact tone, in which there was no petulance. He was seriously alarmed. Perhaps he had a melancholy satisfaction in the prospect of the evil hour being much worse than he had foreseen.

The hour, however, had not yet struck. Suddenly I realised that we were aground. Our arrival was without violence. As placidly as an express train slips into King's Cross a few minutes after covering full sixty miles an hour, our boat ran up against a shelving bank. I leapt ashore, and renewed my attentions to the whustler. He, too, seemed to realise that the battle had entered into new conditions. He bored about, calmly, almost in a weak manner, as if he were a conger-eel. I reeled the line in, and let it out, according to his comings and goings; but I did not stand still. I had to run about a good deal, and in breaking through the scrub, which came down to the edge of the water, was sorely gashed in hands and face and clothes. Never-

theless, my spirits had gone up with a
bound. Even if I lost the whustler, it
was now certain that I should have
nothing to be ashamed of in the morning.
Besides, the squall had gone as suddenly
as it had come. A swell as if of the sea
was swishing on the shore; but there was
not so much as a puff of air, and behind
a vast mass of blackness which I took to
be a shoulder of Ben Ledi there was a
slowly-rising radiance not unlike the
glow that a far-off fire sends upwards to
the clouds of London. Soon the source
of the majestic illumination appeared
above the high horizon. She was
covered and uncovered as the wrack
floated over her face; but she was a
welcome visitor, tempting to gaiety.

"Methinks the moon frowns with a
watery look," said I, inaccurately endeavouring to recall a snatch of appropriate poesy.

"For Goadsake, sir, dinna' sweer—at
this time o' nicht and in a graveyaird!"

"A graveyard?"

"Ay," said Ronald. "D'ye no' ken whaur ye are? Ye're no' on ordnar' warldly land at a'. Ye're on a sma' island, the buryin'-grun' of the Stewarts of Glenbuckie for mair centuries than onybody can remember."

"This is the Inch, then?"

"The same. And no' a canny place ava'. There's naething but wraiths here, and I'll be glad when we're weel awa' frae 't. Hoo's the salmon, sir?"

"Very well, thank you, Ronald. We might get him now if we had a gaff. Just step into the boat and ask the Minister to lend us his."

Ronald obeyed with alacrity. He had not far to go. This being the Inch, we were only two or three hundred yards from the north-east corner of the loch, and not much more from the Kirkton, a hamlet close by the manse.

The boat gone, the whustler had a chance. If only he had made a rush outwards, he could have snapped the tackle and been free. He did not think

of that.   Instead, he sauntered to and
fro, now and then raising himself so high
that I could see his tail slowly waving
above the water in the moonlight.   It
waved sedately, and seemed to be the tail
of a tired whustler ; but I had no bigotry
on that score.   Once, by way of rehears-
ing the final act, which was to go off in
acclaim when Ronald brought the gaff,
I tried to persuade him to come ashore.
I was not successful.   Although the rod
bent into a semicircle, the whustler paid
no heed.   He went on his leisurely way
as if nothing at all were happening.   I
had an uneasy thought that he was
recruiting his energies in contemplation
of a new campaign, and I longed for the
return of the boat.

At length I heard the plash of oars
and the sound of excited voices.   In a
few minutes Ronald and the Minister
came ashore.   I heard the rattle of a
chain, and knew that the boat was being
fastened.

" Hold hard, Ronald," I called out.

"I'm coming aboard whenever I can get him round."

"Takin' her oot to sea again!" said Ronald, aghast. "Mercy on us! what for?"

"To tell you the truth, I don't know. I can't say when we'll get him into the boat; but I am certain we'll never get him into the shore. I've been trying to guide him in; but he won't come. Once or twice he has gone round and round this place, and then it looked as if I were conducting a circus. You wouldn't have me do that all night—in a cemetery, too? Besides, Ronald, if he bolts more than fifty yards we're done, for I can't follow him through the loch on my feet. We're safer in the boat."

"Vera weel, sir," Ronald answered, turning away with a sigh: "I'll bring her roond."

We were now in a situation that required tact, skill, rapidity of judgment and of action. The whustler could not be expected to pause in his stroll for our

convenience. Thus, the boat had to be
"brought round" not a few times, and to
not a few places, before we were safely
seated.

What was to be done next? I
thought it would be well to put off
gently and await the strategy of the
whustler. That came with decision and
energy. Apparently rendered suspicious
by noticing that the slight strain on him
came from a new quarter, he bolted like
a torpedo.   Helped a little by the reel
giving up the line I had recovered,
Ronald made a desperate but successful
effort. The wild rush was soon over.
Trouble, however, was to come. Obeying some strange instinct, the great fish
was making for the Balvaig River, into
which Loch Voil pours its excess. Inwardly I rebuked myself for having left
the comfortable graveyard. There we
might have spent a chill and cheerless
night, with little hope that the dawn
would herald in a brighter day ; but if
we were hauled or lured into the river

the prospect would be nothing less than
disquieting. Had I not read in some
scientific book that salmon travel mainly
by moonlight, and at a speed which the
best of human engines cannot attain?
True, the man of science had been
speaking of salmon when running up the
rivers; but he had not said that when
running down they go with any less
celerity. What, then, if the whustler
got into the Balvaig, which was in
brawling amplitude from nearly a week
of rain? The river has an almost
straight run to the sea. In my startled
imagination I beheld our craft, in tow of
the whustler, leaving Strathyre within
ten minutes; Callander within quarter
of an hour. Rushing past Doune,
ere long we should cross the romantic
Allan Water, and be making full-steam-
ahead for the Firth of Forth. Perhaps
we might look in at St. Margaret's Hope
or at the Port of Leith. There was no
finality to the possibilities with which
the situation was charged. Once in the

North Sea, if we did not turn into Tweed or Tyne, there would be no reason why we should not run up the Thames and make an involuntary appearance before the Terrace of the House of Commons.

It may be that I overestimated the risks suggested by the broad torrent of the Balvaig glittering in the light of the fuliginous moon. I know not. All I know is that when the potentialities of the case burst upon a mind excited by many hours of struggle and high hope I resolved upon an uncompromising measure. Come what might, the whustler must not enter the Balvaig. He must stay in Voil.

"Stop the boat, Ronald," I said, in commanding voice, when, every inch of the line out, I saw the salmon meandering very near a sandbank over which the water of the loch was in motion towards the river.

Then, instead of holding the rod erect, I held it straight out. Followed a game of pull-devil, pull-baker. The real meaning of this phrase was unknown to me;

and even now, recalling the events and
the emotions of that night, I am not
calm enough to be fastidious in philology.
The words seem to express what I wish
to convey, which is that when the
salmon pulled so did I.   Above the
clean yellow sandbank, in which pebbles
were sparkling like diamonds, I saw
him poking, poking, poking; moving
sideways, about a foot at a time, as
if seeking a place at which to dart
across the shallows.   At length he
lost his temper.   Ceasing to struggle
in what may be called a straightforward
manner, he turned a lateral somersault,
and rolled over.   Now, cantrips of that
kind are sometimes an indication that
the game is up, and that practically all is
over but the gaffing.   On this occasion,
however, one had to moderate one's
transports.   I did so by a mental railing
of which I now repent.   " O, William F.
Fisher, of Colorado Springs and the City
of London, why, when you were foolin'
around Noo York, didn't you buy me one

of them tooboolar-steel telescopic poles, calc'lated fit for tarpon, instead of this five-ounce proposition? A Dago, William F.—that's the kind of hairpin You are!" It was touch-and-go with the whustler. Within a time which must have been short though it did not seem so, he rolled himself beyond the point, on the hither side of the sandbank, that was in a straight line with the southern bank of the river, and was once more in the motionless water of the loch. Along the shore he cruised, slowly, silently, and, I think, sadly. He may have been seeking for some definite thing. Ronald and the Minister thought so. On the other hand, he may have been dazed a little, and wandering at random. That was my belief. At any rate, it is not customary for a salmon to move into a brook in spring. That is what the whustler did. Coming to the mouth of a burn not more than three feet wide, he paused a moment as if pondering, and wriggled up.

Ronald pulled the boat ashore, leaped frantically out, squatted down in the mouth of the burn, took a knife from his pocket, and deliberately cut my line.

"Nabbit, nabbit!" he cried. "She's nailed at last!"

"Is he?" I asked, nigh dumb with doubt and amazement.

"Ou, ay," said Ronald in a tone of triumphant certitude. "The Minister couldna' find the gaff—I didna' like to tell ye that a' at aince. But the salmon's richt noo. Ye see, there's a high waterfall no' twenty yairds up among the trees there. She canna' get past that. Neither can she get doon tae the loch again while I sit here, and that I'll do a' nicht. So she'll ha'e to stop in the pool. If the Minister's man will bring me a hay-fork at the scriegh o' day —it winna' be long noo—I'll bring the whustler to the Big Hoose afore breakfast time."

I pondered while lighting my pipe.

Yes: I would allow Ronald to do as he proposed.

On parting for the night the Minister and I arranged to forget about the hayfork. We would be up betimes and go back to the pool unarmed.

INDEX

276 TROUT FISHING

T—— J—— B——, Mr., and his chalk stream, 124, 173 et seq., 218 et seq.
Temperature, the, 54 et seq.
Test, the, 42, 59, 71, 177
Thames, the, 71, 194, 266
Thrush and blackbird, 210
Thunderstorms, 4, 38, 47
Tim the terrier, 152 et seq.
Tod, Mr. E. M., 180
Toryism a matter of taste, 172
Trout, life of the, 157
  feeding, 226
  hearing of, 89
  in April, 140
  miss floating flies, 182 et seq.
  not capricious, 122 et seq.
  order of precedence, 217 et seq.
  seasonal movements of, 223
  sense of smell of, 89
  sleeping, 85
  strength of, 214
Trout's sense of colour, 29
  acute eyesight, 44, 86
  atmospherical sanctuary, 137, 148, 158
Tweed, the, 21, 180, 266
Twilight, 44
Tyne, the, 266

Verbal symbols, 168, 191
Verne, M. Jules, 54
Village sportsmen, 208 et seq.
Voil, Loch, 234 et seq.
  storm on, 256 et seq.

W—— M—— R——, Mr., 178
W—— P——, Sir, 178
Wales, North, 19
Wasps, xi, 89
Water, aeration of, 47, 194
  Cricket, the, 16
  things seen from below, 95 et seq.
Weather, varieties of, 70 et seq.
  illusions about, 98 et seq.
  of British Islands, 109
  prophets, 208
Wemyss, Mr. Erskine, 118
Wey, the, 19
"Whustler," the, 234 et seq.
Wild duck, 88
Winchester, 219
Wind, 37 et seq.
  direction of, 39
  from east or north, 42
  from south or west, 40, 42, 107
  puffs of, 48
  trout feed in high, 52
Winters, old - fashioned, 134
Wordsworth, Mr., 145
Worm-fishing, 213 et seq.
  tackle, 203 et seq.

Yorkshire and Scotland, flies used in, 121

THE END

Printed by R. & R. CLARK, LIMITED, *Edinburgh.*

Printed in Great Britain
by Amazon

50390883R00203